Airborne Radioactive Discharges and Human Health Effects

An introduction

Airborne Radioactive Discharges and Human Health Effects

An introduction

Peter A Bryant
Department of Physics, University of Surrey, UK

IOP Publishing, Bristol, UK

ISBN 978-0-7503-1356-8 (ebook)
ISBN 978-0-7503-1357-5 (print)
ISBN 978-0-7503-1358-2 (mobi)

DOI 10.1088/2053-2563/aafa6d

Version: 20190301

IOP Expanding Physics
ISSN 2053-2563 (online)
ISSN 2054-7315 (print)

British Library Cataloguing-in-Publication Data: A catalogue record for this book is available from the British Library.

Published by IOP Publishing, wholly owned by The Institute of Physics, London

IOP Publishing, Temple Circus, Temple Way, Bristol, BS1 6HG, UK

US Office: IOP Publishing, Inc., 190 North Independence Mall West, Suite 601, Philadelphia, PA 19106, USA

This book is dedicated to all the people who have supported me throughout my life and career. I would like to thank:

Firstly, my parents, Irene and Steve Bryant who have been a continuous source of support throughout my life.

Secondly, my partner Josh Christopher for encouraging me to publish this book.

Lastly my mentors Vince Grindlay, Andy Hicks, Phil Kruse, and Pete Cole who have always been a source of inspiration.

It is hoped this book will provide a useful insight into the risks of radioactive discharges, whether the reader has an interest in the subject or is pursuing a career in this area.

Contents

Preface

In a world post Chernobyl and Fukushima Daiichi, the words 'radioactive discharges' can often be met with concerns of human health effects or impacts to the environment.

A simple search on the Internet on 'health impacts of radioactive discharges' or 'radiation health effects' can find you surrounded by conflicting views around cancers, and what are safe levels of radioactivity.

But what exactly is radiation? Where do airborne radioactive discharges come from? How do they impact the human environment? And how do we regulate these discharges to ensure the protection of the public and environment?

As a professional working in the field of radiation protection for a number of years these are just some of the questions I have been asked.

This book answers these key questions, taking the emotion out of the equation and simply presenting what is internationally accepted scientific consensus.

It starts by providing an overview of the basics of the physics behind radioactivity, from the structure of the atom, to the mechanism of radioactive decay and how radiation interacts with matter.

The book then goes on to look at the sources of airborne radioactive discharges and just how much they impact average background radiation we are exposed to on a daily basis.

Next, the science of how these airborne radioactive discharges move in the physical and human environment is explored along with the concept of radiation 'dose', a measure of radiation exposure and its health impact. This includes a discussion on health effects associated both with routine radioactive discharges, and those discharges from nuclear accidents such as Fukushima Daiichi.

Following on, the techniques for monitoring and sampling airborne radioactive discharges are explored, at both the source of the discharge and in the environment. This includes looking at the measurement of some of the more problematic radionuclides, along with the science behind radiation detection.

The book concludes by focusing on the international conventions, standards and directives in place to control radioactive discharges and how they have been incorporated into the UK domestic regulatory regime. This includes the potential impacts of leaving the European Union and the EURATOM Treaty (BREXIT) on the future of the regulation of radioactive discharges in the UK.

The book is aimed at those with an undergraduate degree in the sciences, whether they are entering a field of work related to radioactivity in the environment, such as the nuclear industry, or those who just have an interest in the area.

It is hoped this book will provide a useful introduction to the science behind radioactivity and radioactive discharges, along with an insight into risks of radioactive discharges.

In the words of Marie Curie:

Nothing in life is to be feared, it is only to be understood. Now is the time to understand more, so that we may fear less.

—Marie Curie

Author biography

Peter A Bryant

A graduate from the University of Bath with a BSc in Physics and University of Surrey with a MSc in Radiation and Environmental Protection, Peter is an experienced professional in Nuclear Safety, Radiation Protection and Radioactive Waste Management. He spent over nine years working in consultancy undertaking projects across the UK and abroad (Lithuania, France and Abu Dhabi). This has included the decommissioning of Chernobyl Style (RBMK) Reactors in Lithuania and the construction of New Nuclear Builds (e.g. Hinkley Point C and Sizewell C).

Peter currently works for EDF Nuclear New Build as a Decommissioning and Radiological Assessment Specialist and Radioactive Waste Adviser. In addition, Peter supports a number of Universities including acting as an Industry Lecturer at the University of Bath, an Associate Tutor at the University of Surrey and Honorary Recognised Teacher at the University of Liverpool.

Peter is also a trustee and President Elect of the UK Chartered Society for Radiological Protection and published in the fields of Radiation Detection, Safety Culture, Skills Development and Risk Assessment. He is a recipient of the Society for Radiological Protection Founders Medal, for his contributions to the field and was the 2018 Young Energy Professional of the Year.

Abbreviations

ALARA	As Low As Reasonably Achievable
BAT	Best Available Techniques
BREXIT	UK exit from the European Union and the EURATOM Treaty
BSS	IAEA International Basic Safety Standards on Radiation Protection and Safety of Radiation Sources
BSSD	EURATOM Basic Safety Standards Directive
CFD	Computational Fluid Dynamics
DEPZ	Detailed Emergency Planning Zone
ERICA	Environmental Risk from Ionising Contaminants: Assessment and Management
EA	Environment Agency
EA(S)R	Environmental Authorisations (Scotland) Regulations 2018
EPR	Environmental Permitting Regulations 2016 (as amended)
EU	European Union
EURATOM	European Atomic Energy Community
GDF	Geological Disposal Facility
HAW	Higher Activity Wastes
HLW	High Level Waste
HSE	Health and Safety Executive
IAEA	International Atomic Energy Agency
ICRP	International Commission on Radiological Protection
ILW	Intermediate Level Waste
IMO	International Maritime Organisation
IRR	Ionising Radiations Regulations 2017
LLW	Low Level Waste
NIA	Nuclear Installations Act 1965
NSL	Nuclear Site Licence
NORM	Naturally Occurring Radioactive Material
NRPB	National Radiological Protection Board
NRW	Natural Resources Wales
ONR	Office for Nuclear Regulations
PWR	Pressurised Water Reactor
REPPIR	Radiation (Emergency Preparedness and Public Information) Regulations 2018
RSR	Radioactive Substances Regulation
SEPA	Scottish Environment Protection Agency
SFAIRP	So Far As Is Reasonably Practicable (UK equivalent to **ALARA**)
TENORM	Technologically Enhanced **NORM**
UNSCEAR	United Nations Scientific Committee on the Effects of Atomic
VLLW	Very Low Level Waste

Glossary of terms

Absorbed dose Measure of energy deposited in any medium, by any type of **ionising radiation**, per unit mass. It has the units Gy or J kg^{-1}.

Active detector Provide real time measurements of radioactivity, dose, or contamination, and as such are most applicable for measuring the activity in a sample.

Aerosol A suspension of solid or liquid particles or particulate in a gas such as air, the most common example of which is dust. The majority of the radionuclides discharged into the atmosphere are in this form and includes vapours such as tritiated water vapours, along with solid particulates such as **radioisotopes** of iodine or cobalt.

Alpha particle Emitted as a result of the radioactive decay of an unstable **isotope**. An alpha particle is equivalent to a positively charged helium (He) nucleus, where the energy of the alpha particle is monoenergetic and unique to that **radioisotope**.

Atom If an amount of an element is repeatedly divided a stage is eventually reached where the element can no longer be sub-divided further. These are called atoms. They are made up of a **nucleus** with orbiting **electrons**.

Atomic number Number of **protons** in an **atom**. This number is characteristic of that **element**.

Background radiation Ionising radiation we are all exposed to from a range of sources in our day-to-day lives.

Bateman equation Mathematical model for describing the activities or abundances in a decay chain as a function of time.

Beta(−) particle Electron emitted as a result of the radioactive decay of an unstable **isotope**. The β(−) particle appears with an energy that varies from decay to decay and can range from zero to the 'endpoint energy'. This endpoint energy is characteristic of the **radioisotope**.

Beta(+) particle Positron emitted as a result of the radioactive decay of an unstable **isotope**. The positron is the anti-particle of the **electron** with the same mass but opposite charge.

Biological half life A term used to measure the rate of excretion of a radionuclide from the human body.

Cloudshine The exposure an individual receives due to radiation from the immersion in the 'cloud' or plume of radioactive aerosols and gases.

Deterministic effects Acute, high dose exposures can result in deterministic effects, these are characterised by a dose threshold, at which an effect is observed and above which the severity of the effect or damage increases with increasing dose.

Diffusion The movement of molecules or **atoms** from a region of high concentration to a region of low concentration as a result of random motion of the molecules or **atoms**.

Dose A measure of radiation exposure, and health impact.

Dose conversion coefficients Coefficients used to convert the uptake of **radioactivity** by a person into an **effective dose** taking account of the radiotoxicity of the radionuclide for a given uptake pathway, as well as biokinetic modelling of the human body.

Dose limitation	Part of the **International System of Radiological Protection**. The fundamental principle is that total dose to any individual from all regulated sources in planned exposure situations, other than medical exposure of patients, should not exceed the appropriate limits recommended by the ICRP.
Effective dose	Takes account of the fact that different organs and tissues within the human body have differing sensitivities to **ionising radiation**. The same units as **equivalent dose**.
Electron	Subatomic particle with a negative charge, significantly smaller than a **proton** or **neutron**. Usually found orbiting the **nucleus** of an atom. Electrons also act as the primary carrier of electricity in a solid.
Element	A substance that cannot be chemically broken down into simpler substances and are primary constituents of all matter.
Emergency exposures	Unexpected situations that occur during operation of a practice requiring urgent action.
Equivalent dose	Takes into account the fact that different types of radiation result a differing degree of damage within a biological system. Units of the Sv, and has the same SI units as absorbed dose of J kg^{-1}.
Existing exposures	Situations that already exist when a decision on control has to be taken, including natural background radiation and residue from past practices.
Exposure pathways	Route by which a person can come in contact with a hazardous substance such as radioactive **aerosol**s or **gases**.
External dose	Radiation exposure where the ionising radiation originates from outside the human body.
Fission	A reaction where the **nucleus** of a heavy nuclei is split into two approximately equal parts, called fission fragments, releasing large quantities of energy.
Foodchain	The various steps in the process by which food is grown or produced and eventually consumed.
Fusion	The process where two light nuclei combine together releasing vast amounts of energy.
Gamma ray	**Photons** produced from radioactive atomic nuclei.
Gas based radiation detector	Built on the principle of an ionisation chamber. A moderate voltage is applied between two electrodes, creating an electric field. **Ionising radiation** that enters the detector may ionise the gas atoms creating an electron ion pair. The flow of ions to the electrodes creates an electric current which is a measure of the radiation in the gas volume.
(Radioactive) Gases	Along with aerosols several of the radioactive discharges are in the form of pure gases. This includes radioactive isotopes of nitrogen, oxygen and carbon dioxide, along with noble gases such as krypton and xenon.
Groundshine	The process of sedimentation can result in radioactive particulates being removed from the plume and deposited on the ground. This has the ability to irradiate a member of the public due to gamma rays emitted by the radioactive particulate distributed in the ground and soil. This is called groundshine.
Half-life	Time taken for half of the nuclei in a sample to decay.

Internal dose	Radiation exposure where the **ionising radiation** originates from inside the human body, for instance due to the inhalation or ingestion of radioactive particulate.
International System of Radiological Protection	Principles used in the protection of people from the potentially harmful effects of ionising radiation. It consists of three principles **justification**, **optimisation** and **dose limitation**.
Intervention	A human activity that seeks to reduce existing radiation exposures or reduce existing likelihood of exposure which is not part of a practice.
Ionising radiation	Electromagnetic waves or subatomic particles with sufficient energy to directly or indirectly knock **electron**s out from an **atom** or molecule, producing ions.
Isokinetic sampling	Sampling at such a rate that the velocity and the direction of the aerosols entering the sampling nozzle is the same as that of the aerosols in the duct/stack. This ensures a representative sample is collected.
Isotope	An **element** with a set number of **proton**s may have a different number of **neutron**s, and as such a different **mass number**. These atoms are called isotopes.
Justification	Part of the **International System of Radiological Protection**. The fundamental principle is that any decision that alters a radiation exposure situation should do more good than harm.
Mass number	Number of **proton**s and **neutron**s in an atomic **nucleus**.
Neutron	Subatomic particle of a similar size to a proton but without an electric charge. Found inside the atomic **nucleus** of an atom. Neutrons are also produced as a form of **ionising radiation**, primarily as a by-product of nuclear reactions.
Non-ionising radiation	Electromagnetic waves or subatomic particles which do not possess sufficient energy to knock out **electron**s from an **atom** or molecule.
Nucleus	The positive charge at the centre of an atom, made up of **proton**s and **neutron**s.
Optimisation	Part of the **International System of Radiological Protection**. The fundamental principle is that the likelihood of incurring exposures, the number of people exposed, and the magnitude of their individual doses should all be kept **ALARA**, taking into account economic and societal factors. In the UK this is implemented through the principles of **BAT** for environmental discharges, and **SFAIRP** for occupational exposures (exposures to workers).
Out of scope	**Radioactivity** limits for solid materials and waste below which the material/waste does not fall under the legal definition of radioactive.
Passive detector	Does not provide a real time measurement of the radiation and tends to be used for measuring **external dose** to an individual for instance a personal dosimeter.
Photon	A particle representing a quantum of light or other electromagnetic radiation.
Planned exposures	Everyday situations involving the planned operation of **practice**s (such as planned discharges of radioactivity).
Practice	A human activity that increases radiation exposure over and above that incurred from background or increases the likelihood of incurring an exposure.

Proton	Subatomic particle with a positive charge equal in magnitude to that of an **electron**. Usually found in the **nucleus** of an atom.
Radioactive decay	Process by which the atomic **nucleus** of a radioisotope loses energy by emitting **ionising radiation**.
Radioactive decay chain	The **radioactive decay** of a **radioisotope** may result in the production of another **radioisotope**, leading to the formation of a decay chain.
Radioactivity	The radioactivity of a radioisotope is a measure of the number of atoms that decay per unit time. It is usually measured in the units of a Becquerel (Bq), where 1 Bq is equivalent to one nuclear disintegration per second.
Radioisotope	An isotope of an atom can be unstable; these are called radioisotopes or radionuclides. Radioisotopes emit ionising radiation via the process of **radioactive decay** to improve their stability.
Representative person	This is defined as an 'individual receiving a dose that is representative of the more highly exposed individuals in the population'.
Safety culture	The assembly of characteristics and attitudes in organisations and individuals which establishes that, as an overriding priority, protection and safety issues receive the attention warranted by their significance.
Sedimentation	Act of gravity on an aerosol, affecting the vertical motion of the particulate.
Segre chart	Chart arranging all the known stable and radioactive **isotope**s by **neutron** number and **proton** number.
Scintillation radiation detectors	These are based on the principle that they emit light (fluoresce) when ionising radiation interacts with them. The light is detected and converted into electrical pulses that are subsequently amplified. The size of the pulse is proportional to the energy deposited in the scintillator material by the incident radiation.
Semiconductor radiation detector	Semiconductor materials exhibit measurable effects when exposed to ionising radiation. By measuring the number of electron–hole pairs, and size of the pulse produced the intensity and energy of the incident radiation can be determined.
Spectroscopy systems	Radiation detection systems used to identify the radionuclides from which the ionising radiation incident on the detector originated.
Source term	The amount of radioactive material released into the environment is called the 'source term' and is presented as a breakdown of the radioactivity (in Bq) by nuclide, normally along with details of the physiochemical form of the discharge and release duration.
Stochastic effects	Chronic, low dose exposures can result in stochastic or probabilistic effects. Unlike deterministic effects there is no threshold associated with stochastic effects, and it is the probability of an effect occurring that increases with dose, rather than the severity of the effect.
Strong nuclear force	Force holding the protons and neutrons of an atomic **nucleus** together.
X-ray	X-rays are photons created from **electron** excitation or acceleration, including those produced by man-made machines.

IOP Publishing

Airborne Radioactive Discharges and Human Health Effects
An introduction
Peter A Bryant

Chapter 1

Radiation physics and the structure of matter

1.1 The structure of the atom

To understand radiation and where it comes from we must begin by looking at the structure of matter. All matter, whether solid, liquid or gas, consists of elements. There 118 known elements as of the end of 2016, of which 92 are naturally occurring, such as hydrogen, carbon and oxygen. The remainder have been produced artificially, including technetium, americium and plutonium.

Elements cannot be broken down using ordinary chemical methods, and as such are defined chemically as the simplest substances in nature [1]. They tend to be chemically linked or combined with other elements, for instance hydrogen and oxygen combine to form water (H_2O). When two or more elements chemically combine, they are known as compounds.

If an amount of an element is repeatedly divided, a stage is eventually reached where the element can no longer be sub-divided further [2]. These individual particles of matter, whose existence was postulated by the Greeks, are called atoms.

Atoms are made up of a nucleus consisting of protons and neutrons, and electrons in shells orbiting around the nucleus, as seen in figure 1.1.

Protons are positively charged, and since like charges repel, electrostatic forces tend to push protons away from each other. Neutrons, with no electrical charge, provide the attractive force (strong nuclear force) that hold the nucleus together.

Electrons are negatively charged, with each electron having a charge of equal magnitude to a proton. An atom normally has an equal number of protons and electrons resulting in the overall neutral charge.

1.2 Mass, atomic numbers and isotopes

Protons and neutrons have an approximately equivalent mass, of a single atomic mass unit (u) or 1.7×10^{-27} kg. An electron is substantially smaller in size than a proton with a mass of 1/1840 u. For this reason, the mass of an atom is determined

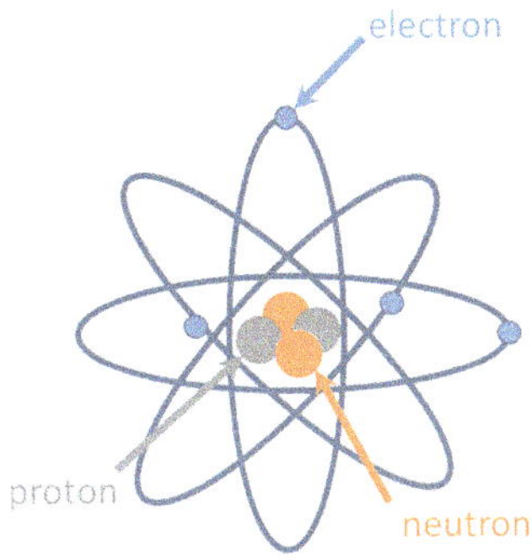

Figure 1.1. The structure of an atom.

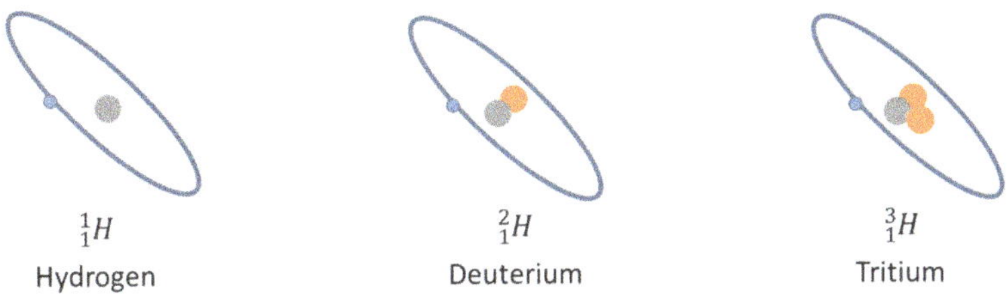

Figure 1.2. Isotopes of hydrogen.

primarily by the number of protons and neutrons. The number of protons and neutrons in the nucleus determines the element's mass number (A).

The number of protons in an atom is known as the atomic number (Z) and is characteristic of that element, for instance hydrogen has one proton, carbon has six protons and oxygen has eight protons. The atoms of a particular element all have the same number of protons.

The notation for a single atom of an element with a chemical symbol X, is written as follows:

$$^A_Z X$$

An element with a set number of protons may have a different number of neutrons, and as such a different mass number. These atoms are called isotopes. All the isotopes of a given element are chemically identical and an example is shown in figure 1.2.

In certain cases, an isotope of an atom can be unstable, these are called radioisotopes or radionuclides. Radioisotopes naturally want to form a more stable nucleus, and emit ionising radiation via the process of radioactive decay to assist in achieving this goal.

1.3 Ionising radiation

We often hear the word 'radiation' mentioned in relation to incidents such as Chernobyl and Fukushima, but it is important to note that what is being referred to is ionising radiation.

Radiation is broadly defined as the emission of energy in the form of waves or particles. This includes light, heat and sound. It is divided into two key types:

- **Ionising radiation**. This is electromagnetic waves or subatomic particles with sufficient energy to directly or indirectly knock electrons out from an atom or molecule, producing ions. This is summarised in figure 1.3.
- **Non-ionising radiation**. This is electromagnetic waves or subatomic particles which do not possess sufficient energy to knock out electrons from an atom or molecule.

As mentioned earlier, radioisotopes emit ionising radiation to form a more stable nucleus. The emission of ionising radiation may result in a change in the number of protons, neutrons or electrons in the radioisotopes, leading to changes in the element and or isotope, alternatively it may just result in a loss of excess energy.

Ionising radiation can be produced by naturally occurring or artificially created radioisotopes, along with being generated by artificial means such as x-ray generators. However, in this book we are primarily interested in naturally occurring and artificially created radioisotopes associated with radioactive discharges as detailed in chapter 2.

Ionising radiation can be further divided as follows:

- **Charged particles**. These strongly interact with matter due to coulombic forces between the charged particles and the positively charged nucleus and negatively charged electrons with the incident atoms. This results in the direct ionisation of the atoms. It includes:
  - alpha particles (α), and
  - beta particles (β).
- **Uncharged particles/photons**. These interact less than charged particles, and tend to ionise indirectly, namely it is secondary interactions that result in the ionisation rather than the incident uncharged particle or photon. This is discussed further in section 1.7. It includes:

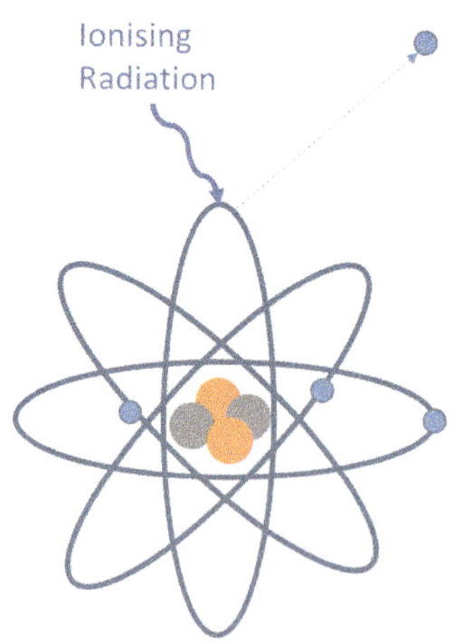

Figure 1.3. Ionisation of an atom.

- x- and gamma-rays (γ), and
- neutrons (n).

1.4 Mechanism of radioactive decay

Radioactive decay is a random event, and as such it is impossible to predict the instant in time when such a decay will occur.

The activity (A) of a radioisotope is a measure of the number of atoms that decay per unit time. It is usually measured in the units of a Becquerel (Bq), where 1 Bq is equivalent to one nuclear disintegration per second [3]. The activity can be expressed using the differential form:

$$A = -\frac{dN}{dt} = \lambda N \tag{1.1}$$

Where N is the number of radioactive atoms, t is time and λ is the decay constant, which is specific to a radioisotope. Integrating equation (1.1) gives the radioactive decay law which is expressed mathematically as follows:

$$N_t = N_0 e^{-\lambda t} \tag{1.2}$$

Where N_0 is the number of nuclei present initially and N_t is the number of nuclei present at time t. Since the activity of the sample is proportional to the number of radioactive atoms, this also varies exponentially:

$$A_t = A_0 e^{-\lambda t} \tag{1.3}$$

The rate at which a radioactive isotope decays is commonly measured using the term half-life (t_r), this is the time taken for half of the nuclei in a sample to decay. This is demonstrated in figure 1.4, where the blue atoms represent the parent radioactive isotope and the black atoms represent the stable progeny atoms.

The half-life is specific to a radioisotope and directly related to the decay constant, λ as follows:

$$t_r = \frac{\ln 2}{\lambda} \tag{1.4}$$

1.5 Decay chains and equilibrium

The radioactive decay of a radioisotope may result in the production of another radioisotope. This results in a decay chain, as seen in figure 1.5.

The activities and abundances of radioactive isotopes within a decay chain are not independent. They are determined by the history of the decay, e.g. the decay rates and abundances in the preceding part of the chain.

The Bateman equation is a mathematical model for describing the activities or abundances in a decay chain as a function of time. It is not a single equation, but rather a series of differential equations describing the chain of interest.

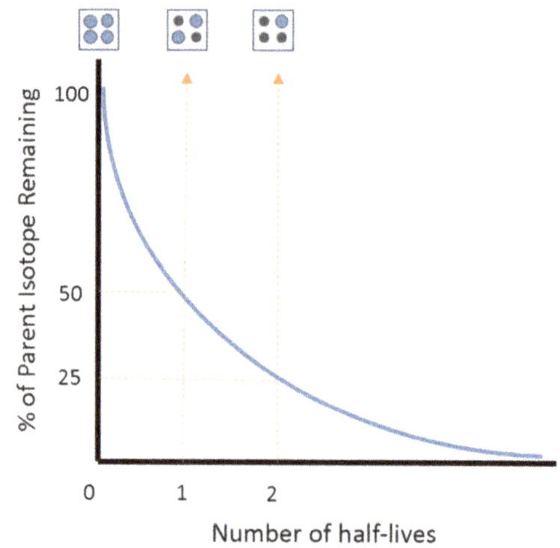

Figure 1.4. Diagram showing the change in activity of an isotope over time.

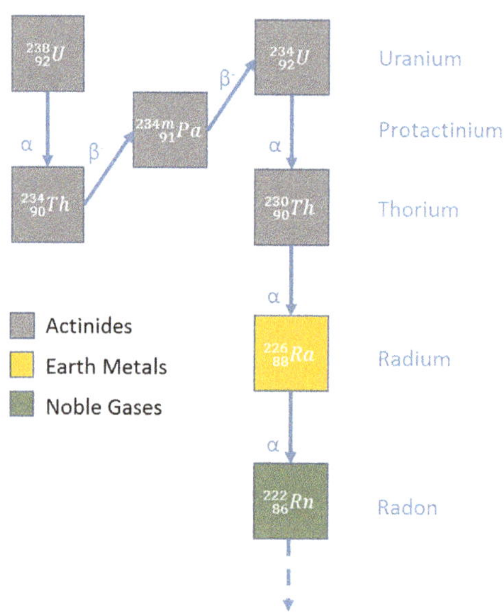

Figure 1.5. Portion of the decay chain for uranium-238.

Consider a decay chain consisting of two sets of decays, where the parent radioisotope decays into a progeny radioisotope, which subsequently decays into a stable isotope. The activity of the parent radioisotope can be modelled using a version of equation (1.1) or equation (1.5), whereas the equation for the progeny radioisotope needs to include a term describing the decay of the progeny to a stable isotope, but also a term describing the creation of the progeny radioisotope by the parent. This is summarised in equations (1.5) and (1.6):

$$\frac{dN_P}{dt} = -\lambda_P N_P \tag{1.5}$$

$$\frac{dN_D}{dt} = -\lambda_D N_D + \lambda_P N_P \tag{1.6}$$

where N_P and N_D are the number of parent and progeny atoms at time t, and λ_P and λ_D are the parent and progeny decay constants.

The solution to the Bateman equation, allows us to define three useful situations.

- **Transient equilibrium**. Where the half-life of the of progeny radioisotope is less than the parent radioisotope ($\lambda_D > \lambda_P$) an equilibrium is reached by the parent–progeny radioisotope pair. In this case the number of progeny atoms can be calculated using the following equation:

$$N_D = \frac{\lambda_P N_P}{\lambda_D - \lambda_P} \tag{1.7}$$

- **Secular equilibrium**. In the event the half-life of the progeny radioisotope is negligible (much shorter) compared to the parent radioisotope ($\lambda_D \gg \lambda_P$) the activity or abundance of the radioactive isotope remains constant as its production rate is equal to its decay rate. Namely:

$$A_D = A_P \tag{1.8}$$

- **No equilibrium**. Should the half-life of the progeny radioisotope be greater than the parent radioisotope ($\lambda_D < \lambda_P$) then equilibrium will never be reached. In this event calculating the number of progeny atoms is more complicated, as shown in equation (1.9)

$$N_D = \frac{\lambda_P N_P^0 (e^{-\lambda_P t} - e^{-\lambda_D t})}{\lambda_D - \lambda_P} + N_D^0 e^{-\lambda_D t} \tag{1.9}$$

where N_P^0 and N_D^0 are the initial number of parent and progeny atoms at time 0.

1.6 The nuclide chart

The Segré chart or nuclide chart arranges all the known stable and radioactive isotopes by neutron number and proton number. An overview of the main characteristics of the Segré chart are summarised in figure 1.6.

The black line represents the stable isotopes or nuclides, whereas the orange region represents the unstable radioactive isotopes.

As mentioned previously, radioisotopes emit ionising radiation leading to a change in the number of protons, neutrons or electrons in the radioisotope as it moves towards the black line (stable isotopes).

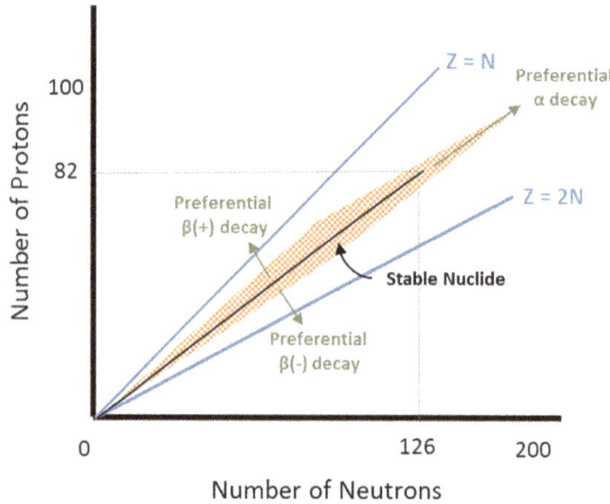

Figure 1.6. Segré chart.

Below the black line radioisotopes primarily decay by $\beta(-)$ decay, whereas above the black line radioisotopes primarily decay by $\beta(+)$ decay. Above a proton number of 82, radioisotopes tend to decay by α decay. This is discussed further in section 1.7.

In most cases, after the emission of an α or β particle the nucleus of the radioisotope rearranges itself slightly, releasing energy in the form a γ-ray.

1.7 Types of radioactive decay

As discussed in section 1.3, ionising radiation can take the form of four main emission types, alpha, beta, x- and gamma-rays, and neutrons. Each of these decay types is discussed further in the following sub-sections:

1.7.1 Alpha decay and the alpha particle

As seen in figure 1.6, above an atomic number of 82 isotopes tend to be unstable and radioactive. The easiest way for these isotopes to remove the excess numbers of protons and neutrons is to eject them directly from the nucleus in the form of an α particle. The decay process is written as:

$$^{A}_{Z}X \rightarrow {}^{A-4}_{Z-2}Y + {}^{4}_{2}\alpha$$

An α particle is equivalent to a positively charged helium (He) nucleus and its ejection is a spontaneous process where the energy of the alpha particle is monoenergetic (typically of the order of a MeV or 10^{-13} J) and unique to that radioisotope. This is summarised in figure 1.7, which shows an example of typical alpha spectra obtained from a radiation detector.

The broadening of the peak is an effect of the detection process, due to the alpha particle depositing all its energy at different depths within the detector, rather than the alpha particle having a range of energies.

When an α particle passes through matter it strongly ionises, losing its energy quickly. Due to its positive charge, it repels atomic nuclei and attracts atomic

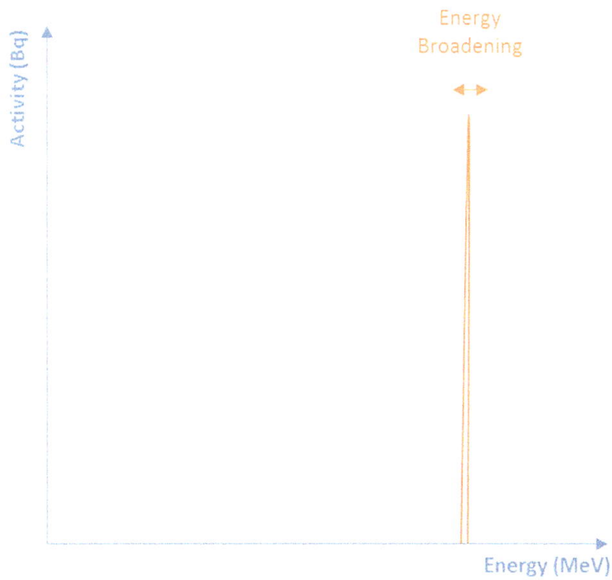

Figure 1.7. Typical α spectrum.

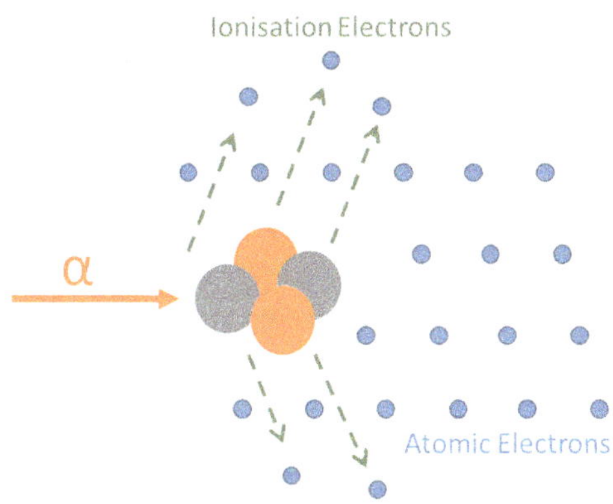

Figure 1.8. α particle interacting with matter.

electrons; furthermore, owing to its large size relative to the electron it can interact with several atomic electrons at once. This results in a lot of ionisation but also means that the alpha particle loses its energy very quickly and does not penetrate far into a material. This is demonstrated in figure 1.8.

1.7.2 Beta decay and the beta particle

For neutron rich radioactive isotopes with an atomic number less than 82, the nucleus does not eject the neutrons but, rather may try and reach stability by converting a neutron into a proton, by emitting a high-speed electron or $\beta(-)$ particle, as it is more commonly known.

This type of decay not only happens at lower atomic numbers but also may occur due to the formation of an unstable neutron rich isotope from an α decay, as seen in the decay chain in figure 1.5.

The $\beta(-)$ decay process is written as:

$$^{A}_{Z}X \rightarrow {}^{A}_{Z+1}Y + \beta(-) + \bar{\nu}$$

The $\beta(-)$ particle appears with an energy that varies from decay to decay and can range from zero to the 'endpoint energy', as seen in figure 1.9. This endpoint energy is characteristic of the radioisotope.

In order to ensure conservation of energy, an antineutrino ($\bar{\nu}$) is emitted along with the $\beta(-)$ particle, such that the total energy of the $\beta(-)$ particle and $\bar{\nu}$ is always equal to the endpoint energy for that radioisotope. The $\bar{\nu}$ has a mass of ~0 and is non-charged. It is therefore very difficult to detect and rarely interacts with matter.

Due to its size, the $\beta(-)$ particle is less ionising than an α particle, however as it is charged it also interacts with matter via coulombic forces. When a $\beta(-)$ particle interacts with an atomic electron it scatters off the electron due to the repulsion of their electrical charges. Should sufficient energy be passed to the atomic electron during this interaction it will cause ionisation, as seen in figure 1.10.

A $\beta(-)$ particle may also interact with the nucleus of an atom due to its large positive charge. When a $\beta(-)$ particle passes close to a highly charged nucleus (found in materials with higher atomic numbers), its trajectory may become tightly bent by the coulombic force, as seen in figure 1.11. This causes the $\beta(-)$ particle to lose momentum, and the loss of kinetic energy converted into the emission of x-rays. These x-rays are called bremsstrahlung radiation. Because of bremsstrahlung, low density materials such as plastics are normally used when shielding against $\beta(-)$ radiation.

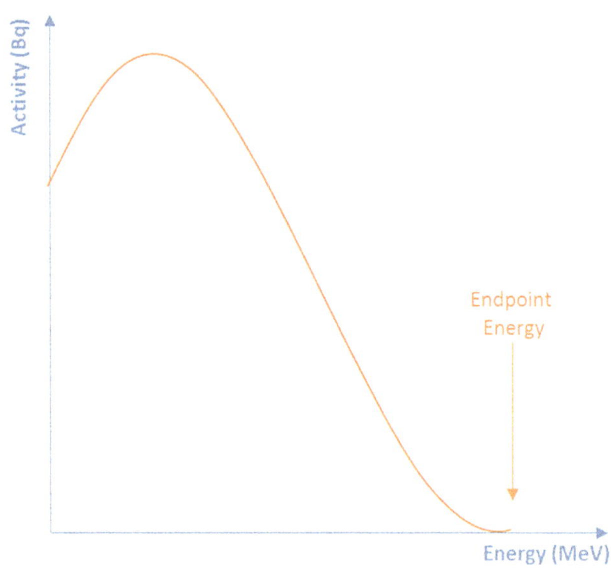

Figure 1.9. Typical β spectrum.

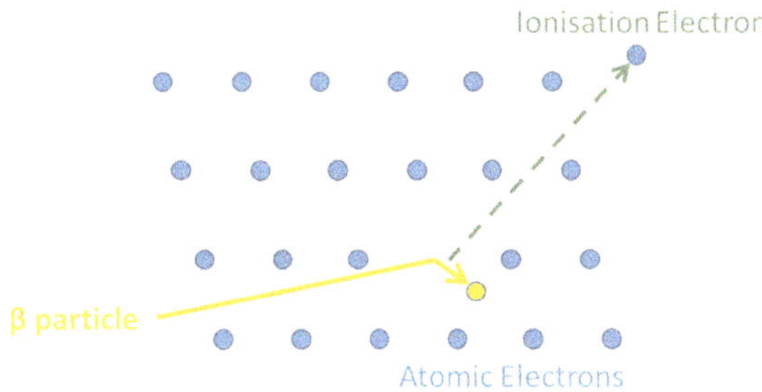

Figure 1.10. $\beta(-)$ particle interacting with atomic electrons.

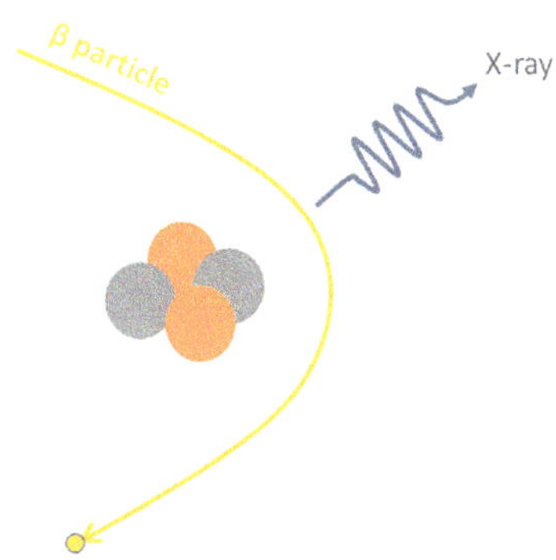

Figure 1.11. $\beta(-)$ particle interacting with an atomic nucleus.

In a similar fashion to the $\beta(-)$ decay, unstable proton-rich isotopes may convert an excess proton to a neutron in an attempt to reach stability. This may happen by one of two means.

- **Electron capture**. This is where a proton coverts into a neutron by capturing one of the nearest orbital electrons. The excess energy from capturing the orbital electron is converted into a neutrino. Like the antineutrino, the neutrino is very difficult to detect and rarely interacts with matter.

$$ {}^{A}_{Z}X \rightarrow {}^{A}_{Z-1}Y + v $$

- **$\beta(+)$ decay**. If electron capture is not possible, the proton can be converted into a neutron by the emission of a positron, or $\beta(+)$ particle; this is identical to an electron in size but with a positive charge

$$ {}^{A}_{Z}X \rightarrow {}^{A}_{Z-1}Y + \beta(+) + v $$

When the positron ($\beta+$) is in the presence of matter it does not exist for a long period before it combines with an electron, and annihilates. During this process the masses of both particles are converted into two 511 keV gamma-rays which are emitted in opposite directions. These gamma-rays go on to cause further ionisation. This is summarised in figure 1.12.

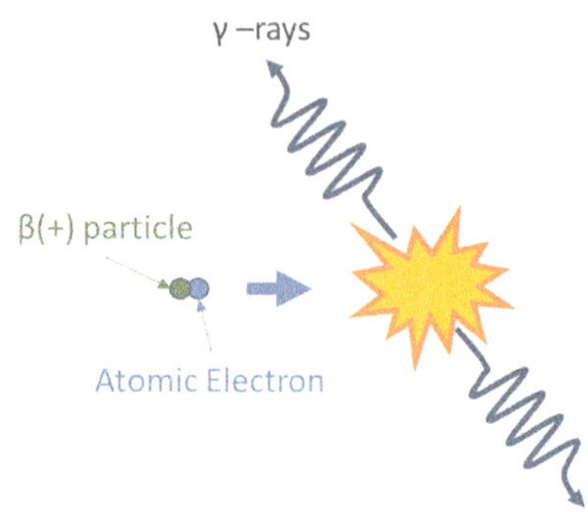

Figure 1.12. $\beta(+)$ particle interacting with an atomic electron.

1.7.3 X- and gamma-rays

X- and γ-rays are both high-energy electromagnetic waves. They are broadly distinguished by the processes that made them.

X-rays are photons created from electron excitation or acceleration, including those produced by man-made machines.

As mentioned earlier, after the emission of an α or β particle the nucleus of the radioisotope rearranges itself slightly, releasing energy in the form of a γ-ray. For this reason, γ-rays are often defined as photons produced from radioactive atomic nuclei. This process is written schematically as:

$$ {}^{A}_{Z}X \rightarrow {}^{A}_{Z}X + \gamma $$

It should be noted that the emitted γ-ray is unique to that radioisotope. Figure 1.13 shows a typical γ-ray spectrum, the photopeak is the unique energy of the γ-ray or

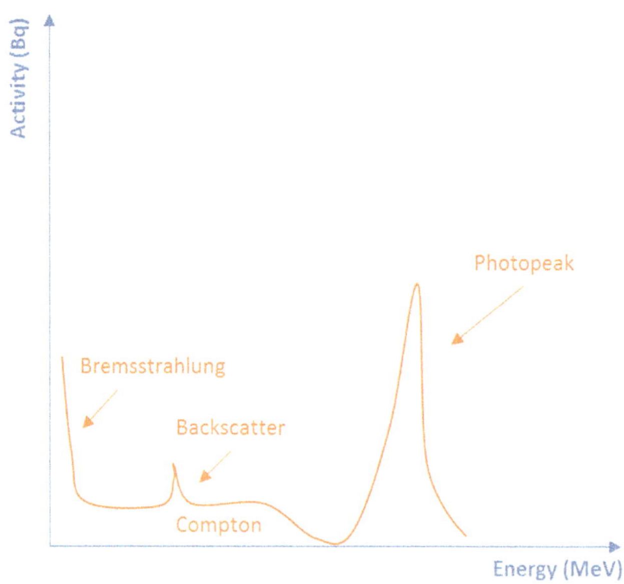

Figure 1.13. Typical γ-ray spectrum.

γ-rays for that radioisotope. The peak broadening and artefacts in the spectra are related to the γ-ray interactions with the detector material. This is discussed further later in this sub-section and chapter 6.

Both x-rays and γ-rays are less ionising than charged particles, due to the lack of coulombic forces, and as such can penetrate further through matter. They ionise indirectly via three different mechanisms, depending on their energy.

- **Photoelectric ionisation**. X-rays and lower energy γ-rays (< 0.5 MeV) may be absorbed by an atomic electron. If this energy is great enough, the electron will be ejected from the atom. The ejected electron then goes on to to cause further ionisation in the matter, in the fashion described for $\beta(-)$ particles. It should be noted that the entire energy of the incident photon is absorbed by the atomic electron during photoelectric ionisation (figure 1.14).
- **Compton scattering**. For medium energy x- and γ-rays (0.5–1 MeV) the atomic electron might not absorb all the energy of the incident photon. This results in the photon scattering off the electron, losing some of its energy, before going on to cause further ionisation (figure 1.15). It is this interaction mechanism that creates the Compton region and backscatter peak in a γ-ray spectrum, as seen in figure 1.13.
- **Pair production**. High-energy γ-rays (> 1.022 MeV) can interact with the atomic nuclei in a material. This can result in the production of an electron

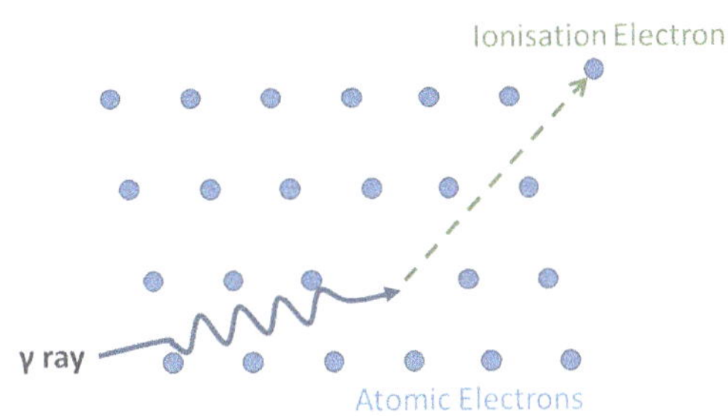

Figure 1.14. Photoelectric ionisation.

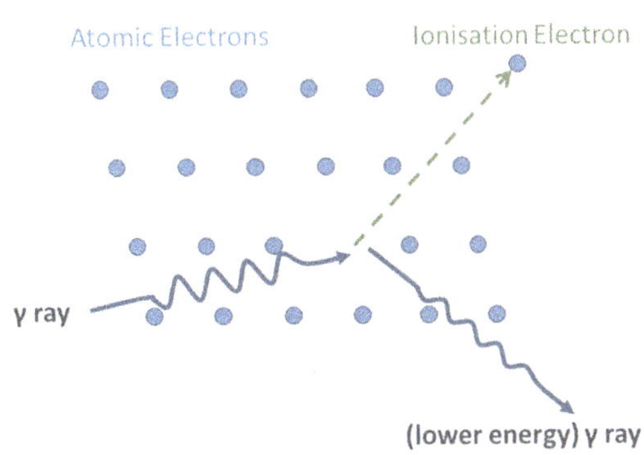

Figure 1.15. Compton scattering.

and positron pair. The electron and positron then go on to cause further ionisation much like a $\beta(-)$ and $\beta(+)$ particle (figure 1.16).

1.7.4 Neutron sources and interactions with matter

Unlike alpha, beta and gamma radiation, which is primarily generated by unstable isotopes, neutrons are mainly produced as a by-product of nuclear reactions. For this reason, neutron sources are not covered in detail within this book, as they do not normally form part of radioactive discharges. An example of one of the more common nuclear reactions used to produce neutrons is shown below, along with the associated neutron spectrum (figure 1.17).

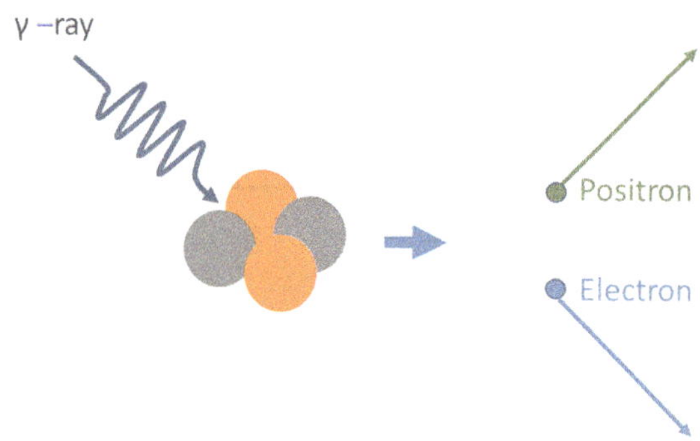

Figure 1.16. Pair production.

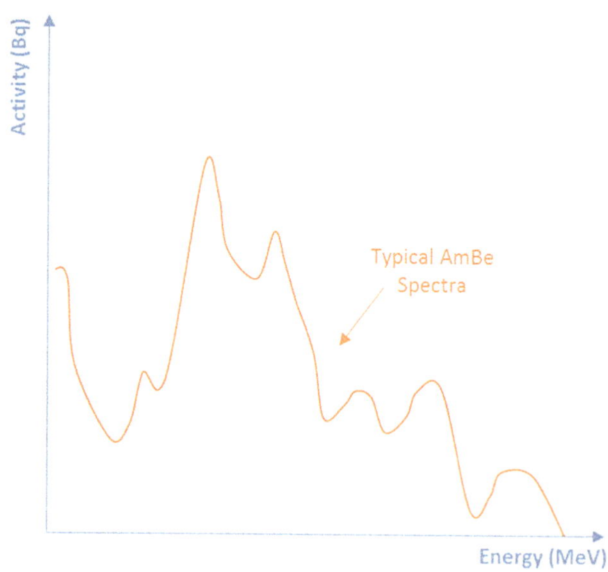

Figure 1.17. Typical americium-241 beryllium (AmBe) neutron spectrum.

$$^{4}_{2}\alpha + ^{9}_{4}Be \rightarrow ^{12}_{6}C + ^{1}_{0}n$$

Neutrons are also highly penetrating but cause much more ionisation than x-rays and γ-rays (but less than directly ionising radiation e.g. α and β particles).

Neutrons ultimately transfer energy to atomic nuclei, this can either be due to scattering, which may be elastic or inelastic, or it can undergo nuclear interactions with the atomic nuclei in a material.

After interacting with a neutron, the nucleus may then emit gamma radiation, it may recoil from its position and cause ionisation in a similar way to an alpha particle or can fission. These interaction mechanisms are summarised in figure 1.18.

1.8 Penetrating powers of ionising radiation

As discussed in section 1.7, α and β particles, interact with matter via coulombic forces, resulting in high levels of ionisation, but also the radiation losing its energy over a relatively short distance. This means that radioisotopes that emit α and β particles tend to pose the greatest hazard if they are ingested, inhaled or injected into the human body, where they can cause ionisation and disrupt the delicate chemical processes of the body, and even cause cell death.

External to the human body α and β particles are unlikely to penetrate past the skin, and therefore the potential health impacts are minimal.

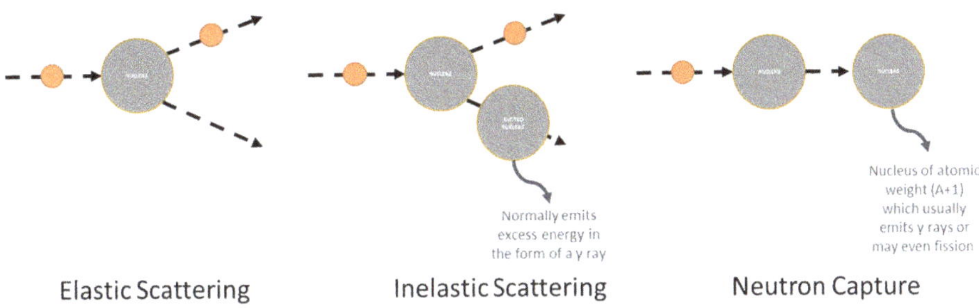

Figure 1.18. Typical neutron interactions.

Table 1.1. Properties of ionising radiation (adapted from [4]).

Radiation	Form	Relative mass	Charge	Range in air	Range in human tissue
Alpha particle	2 Protons and 2 neutrons	4	+2	~3 cm	~0.04 mm
Beta particle	Electron/positron	1/1840	−1 or +1	~3 m	~5 mm
X- and gamma-rays	Electromagnetic radiation	0	0	Large	Through body
Neutrons	Neutron	1	0	Large	Through body

In the case of x- and γ-rays and neutrons, these are less ionising than charged particles and as such penetrate through the body, whether the source of the radiation is internal to the body or external to the body. For this reason, these types of ionising radiations are hazardous when external to the human body.

A summary of the properties of the various types of ionising radiation is presented in table 1.1.

1.9 Summary

- Atoms are made up of a nucleus consisting of positively charge protons and uncharged neutrons, with negatively charged electrons in shells orbiting around the nucleus.
- The number of protons and neutrons in the nucleus determines the element's mass number (A).
- The number of protons in an atom is known as the atomic number (Z) and is characteristic of that element
- An element with a set number of protons may have a different number of neutrons, these atoms are called isotopes.
- An isotope of an atom can be unstable; these are called radioisotopes or radionuclides. Radioisotopes emit ionising radiation via the process of radioactive decay to improve their stability.
- Ionising radiation is electromagnetic waves or subatomic particles with sufficient energy to directly or indirectly knock electrons out from an atom or molecule, producing ions.
- There are four types of ionising radiation:
  - Alpha particles (α)—An α-particle is equivalent to a positively charged helium (He) nucleus. It is strongly ionising, but loses its energy over a short range as it passes through matter.
  - Beta particles (β)—A β particle is a high-speed electron or positron. It is less ionising than an α particle, however, it loses its energy over a greater range.
  - X- and gamma-rays (γ)—An x-ray originates from electron excitation or acceleration, whereas γ-rays are produced from radioactive atomic nuclei. They can penetrate deep into matter but are indirectly ionising, and ionise less than both β and α particles.
  - Neutrons (n)—Neutrons are primarily produced as a by-product of nuclear reactions. They are highly penetrating but cause much more ionisation than x-rays and γ-rays (but less than directly ionising radiation e.g. α and β particles).
- The activity of a radioisotope is usually measured in the units of a Becquerel (Bq), where 1 Bq is equivalent to one nuclear disintegration per second.
- The rate at which a radioactive isotope decays is commonly measured using the term half-life (t_r), this is the time taken for half of the nuclei in a sample to decay.

- The radioactive decay of a radioisotope may result in the production of another radioisotope, leading to the formation of a decay chain.
- The Segré chart arranges all the known stable and radioactive isotopes by neutron number and proton number.
- α and β particles tend to pose the greatest hazard if they are ingested, inhaled or injected into the human body.
- X-/γ-rays and neutrons are hazardous when external to the human body.

References

[1] The Periodic table of Elements Jefferson Lab http://education.jlab.org/itselemental/index_num.html [accessed 12 March 2017]

[2] Sharon M 2009 *Nuclear Chemistry* 1st edn (Boca Raton, FL: CRC Press)

[3] Knoll G F 2000 *Radiation Detection and Measurement* (New York: Wiley)

[4] Martin A, Harbison S, Beach K and Cole P 2012 *An Introduction to Radiation Protection* 6th edn (Boca Raton, FL: CRC Press)

Chapter 2

Sources of radioactive discharges

As discussed in chapter 1, ionising radiation can be produced by naturally occurring or artificially created radioisotopes, along with being generated by artificial means such as x-ray generators.

In our day-to-day lives we are all exposed to ionising radiation from a range of sources, this is called background radiation. The annual exposure to a member of the public from background radiation varies based on location, with an average worldwide value of 2.4 mSv yr^{-1} [1]. Where an mSv is a unit of radiation exposure, which is discussed further in chapter 5.

A breakdown of the various sources of background radiation is summarised in figure 2.1.

On average, the majority of the ionising radiation (~84%) comes from naturally occurring radioisotopes including cosmic radiation from space, radon from the ground, and radioisotopes in building materials, food and water [2]. Artificial or man-made sources of ionising radiation come primarily in the form of medical exposures such as x-rays.

The focus of this book is in relation to radiation exposure to members of the public from gaseous radioactive discharges. This constitutes only a very small portion of the average background exposure (~0.01%). Radioactive discharges originate from several different sources, the most common of which are in relation to the nuclear industry. Other sources include educational establishments, hospitals, waste handling and disposal facilities, and the oil and gas industry.

Further details on these sources of radioactive discharges is provided within this chapter, along with details of the physiochemical form of the discharges.

2.1 Discharges from civil nuclear industry

The most well-known source of radioactive discharges is associated with nuclear reactors. Like other sources of power generation, they work on a very similar principle to a kettle, namely heating water, which in turn produces steam. The steam

doi:10.1088/2053-2563/aafa6dch2

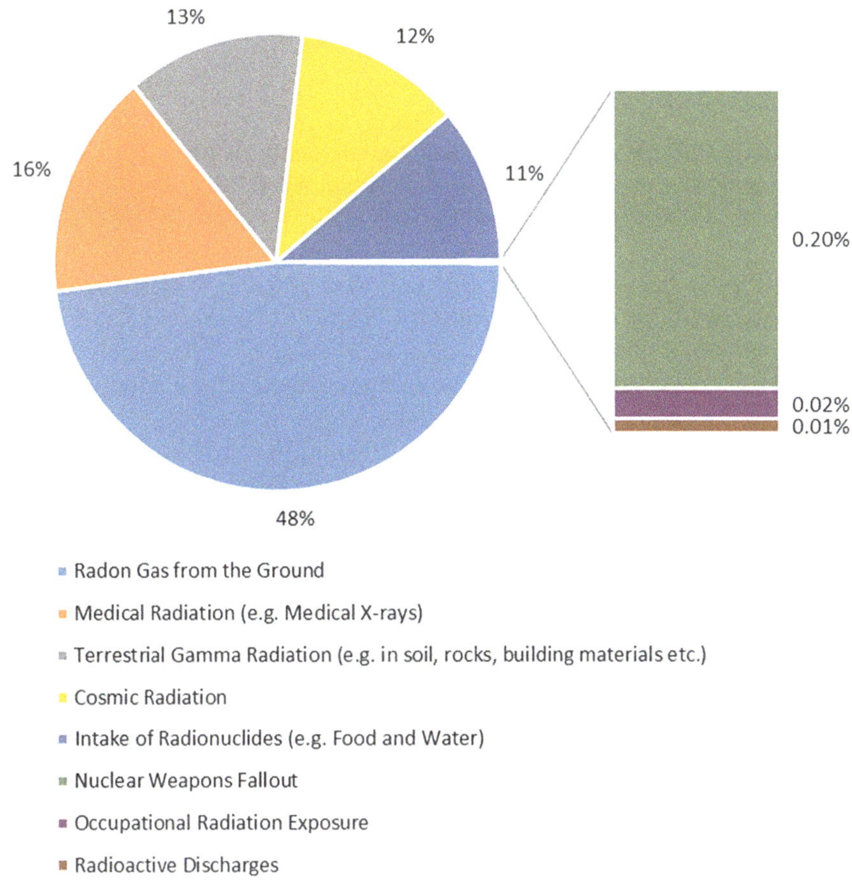

Figure 2.1. Breakdown of the various sources of background radiation.

in turn drives a turbine producing electricity, where the higher the purity of the steam (lower level of contaminants and water vapour droplets) the greater the efficiency of the reactor.

Unlike conventional coal fired power stations which burn coal to generate heat, a nuclear reactor generates heat via a process called nuclear fission. This is a reaction where the nucleus of a heavy nuclei, is split into two approximately equal parts, called fission fragments, releasing large quantities of energy.

Some nuclei undergo spontaneous fission, where the splitting of the nucleus occurs by itself, however, this is normally at a rather low rate, whilst other nuclei may undergo fission when they are bombarded with neutrons [3]. The latter nuclei are called fissile, and examples include uranium-235 (^{235}U) and plutonium-239 (^{239}Pu).

An important part of the fission process, is that the fission fragments are unstable releasing neutrons, along with other forms of ionising radiation. These neutrons go on to interact with other fissile nuclei, causing further fission, resulting in what is known as a chain reaction. This chain reaction is vital to the fission process being a viable energy source.

A summary of the nuclear fission process is provided in figure 2.2.

2.1.1 Components of a nuclear reactor

There are many different types of nuclear reactor designs in the world. One of the most common is the pressurised water reactor or PWR. An overview of the key components in a PWR are presented in figure 2.3, with further details provided below:

- **Fuel**. This is where the fission reaction takes place. It is typically made from enriched uranium, where the ratio of ^{235}U to ^{238}U is around 5%. This ratio is important to ensuring a chain reaction as it is the ^{235}U which is the main source of the fission reaction. The fuel is covered by a cladding material to prevent the escape of the fission products.
- **Moderator**. Neutron absorption (hence fission) in ^{235}U is more likely to occur with slow neutrons (thermal neutrons) than the fast neutrons produced by the fission products. Using a moderator such as water slows the neutrons so they are more likely to interact with other uranium atoms. This means the fission process can go on indefinitely in a chain reaction.
- **Control rods**. Control rods are concentrated neutron absorbers (poisons) which can be moved into or out of the core to change the rate of fission in the reactor. Rod insertion adds neutron poisons to the core area, which makes fewer neutrons available to cause fission which results in a reduction in heat production and power. Pulling the control rods out of the core removes poisons, allowing more neutrons to cause fission and increasing reactor power and heat production.

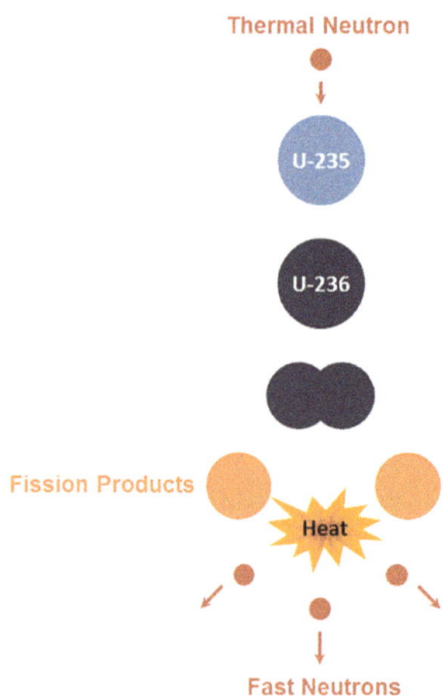

Figure 2.2. Overview of the fission process.

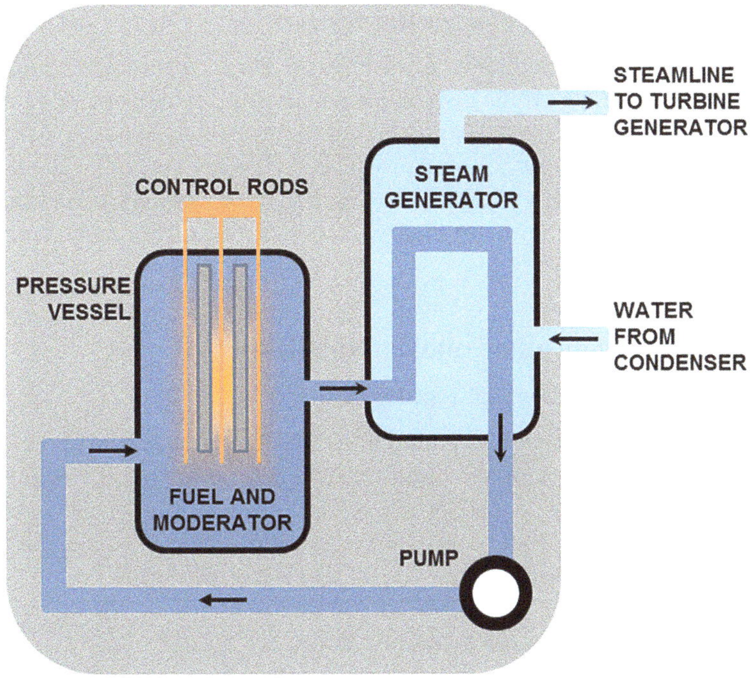

Figure 2.3. Key components in a pressurised water reactor (PWR).

- **Coolant**. Heat from the fuel is removed by the coolant. In a PWR, water serves as both the moderator and coolant, which has many benefits due to temperature effects on moderation. Most coolants become radioactive to some extent, both from the capture of neutrons by nuclei in the coolant (activation), and from leaks of fission fragments from the fuel into the coolant.
- **Steam generator**. Depending on the design of the reactor, steam can be generated either directly by the heating of the coolant, or from steam generators. In a PWR the water coolant is kept under pressure to prevent it boiling and producing steam. Instead the heated coolant is passed through a steam generator, heating water in the secondary loop of the reactor, which in turn generates steam. As the reactor coolant does not come into direct contact with the water in the secondary loop, there is a low risk of radioactive contaminants in the steam, however, there is a loss in the heat energy transferred between the fuel and the steam.
- **Turbine**. The turbine converts the heat within the steam into mechanical energy, which in turn generates electrical power. As mentioned earlier, the higher the purity and temperature of the steam, the greater the efficiency of the conversion process. The presence of contaminants and water vapour droplets can damage the turbine reducing its efficiency and increase maintenance requirements.
- **Condenser** The condenser converts the steam that passes through the turbine back into water, which re-enters the steam generator.

2.1.2 Radioactive discharges from nuclear reactors

Nuclear reactors generate radioactive waste in the form of gaseous, liquid and solid waste, however it is the gaseous discharges which are the focus of this book.

The gaseous discharges arise from the ventilation of radioactive or potentially radioactive areas and process equipment within the power station. The radionuclide breakdown of the discharges varies due to the reactor technology. An example of the typical radionuclides that are discharged from a pressurised water reactor are presented in table 2.1.

2.2 Discharges from other nuclear facilities

2.2.1 Nuclear fuel production and reprocessing

Prior to use in a nuclear reactor, uranium must first be converted into a suitable form to be fissioned. A summary of the uranium fuel cycle is provided in figure 2.4.

Table 2.1. Gaseous radioactive discharges from a typical 1650 MW (electric) PWR.

Radionuclide	Discharge (GBq yr^{-1})
Tritium (^{3}H)	500
Carbon-14 (^{14}C)	350
Noble gases	800
Iodine-131 (^{131}I)	0.025
Fission and activation products	0.004

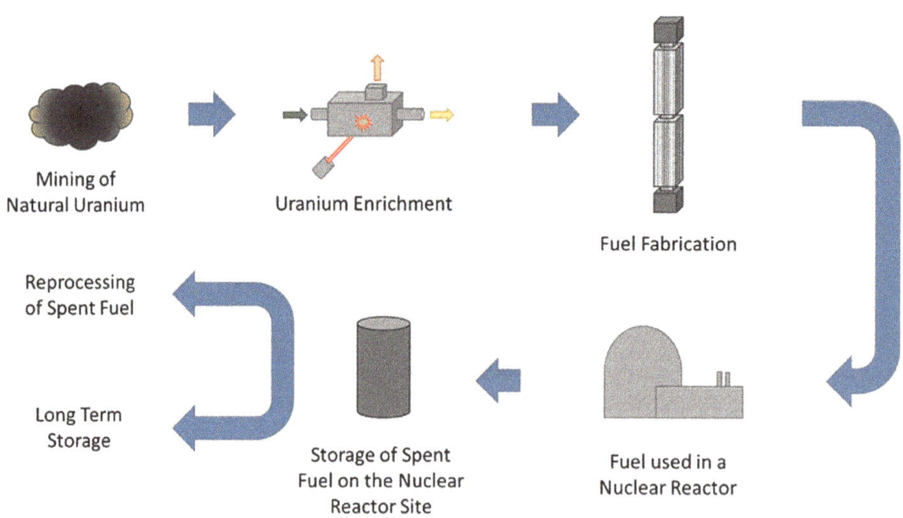

Figure 2.4. Summary of the uranium fuel cycle.

Natural uranium ore is extracted from both underground mines and open-pit mines. The uranium ore is crushed into a fine powder, and following a chemical leaching process produces what is known as yellowcake uranium [4]. The ore and yellowcake uranium has a very low level of radioactivity, so much so that the chemical toxicity of the material tends to be of greater concern than its radioactivity [11]. For this reason, the environmental impacts associated with the mining of uranium are the same as mining any other metalliferous ore, and radioactive discharges are negligible.

As previously mentioned, to sustain a chain reaction within a nuclear reactor, the uranium needs to be enriched. To do this the yellowcake uranium is converted into a gaseous form called uranium hexafluoride (UF_6) [6]. This is then enriched by one of three processes:

- **Diffusion based enrichment**. Like the name suggests, this involves the use of a permeable barrier and the process of diffusion. The molecules of $^{235}UF_6$ move across the barrier at a slightly faster rate than those of the $^{238}UF_6$. This allows the ratio of ^{235}U to ^{238}U to be increased by a small amount. The process must be repeated thousands of times to obtain the required enrichment for power generation.
- **Centrifuge enrichment**. The UF_6 gas is fed into a centrifuge which spins at a high rate, causing the lighter ^{235}U atoms to concentrate near the centre of the centrifuge, due to centripetal forces, while the heavier ^{238}U accumulate along the wall. The mixtures are subsequently siphoned off. Centrifuge enrichment is far more efficient than diffusion based enrichment, both in terms of enrichment rate and energy consumption, however, the construction costs are high.
- **Laser enrichment**. This is a predominantly experimental process using a laser to ionise the ^{235}U atoms within the UF_6 gas. The $^{235}UF_6$ are then separated based on its electrical charge.

Much like a power station, during the enrichment processes gaseous discharges arise from the ventilation of radioactive or potentially radioactive areas and process equipment. The discharges are predominantly uranium radioisotopes, and uranium progeny, with the exact levels of discharges varying depending on the throughput and enrichment technology choice. Typical discharges of uranium radioisotopes from the UK enrichment facilities at Capenhurst, which operates three centrifuge enrichment plants is around 10^6 Bq yr^{-1}.

Following enrichment, the enriched UF^6 is converted into uranium dioxide powder and fabricated into ceramic pellets in a furnace. These are then stacked to form fuel pins, which in turn are assembled into fuel assemblies. As with the enrichment process the levels of discharges, vary depending on throughput and the type of fuel being fabricated (which varies depending on the reactor technology choice). Typical discharges of uranium radioisotopes from the UK fuel fabrication at Springfields, is around 10^7–10^8 Bq yr^{-1} [7, 8].

The fuel assemblies are subsequently used in commercial nuclear reactors. Once the fuel has been 'burned up' and a chain reaction is no longer sustainable, due to the reduced ratio of ^{235}U to ^{238}U and formation of fission products in the fuel, it is usually stored on the site of the nuclear reactor for a period of time to allow the fuel to cool. This is typically for a period of 5–10 years, due to the high levels of heat being generated by the fission fragments within the fuel. Following the cooling period the spent fuel is either put into long-term storage, prior to permanent disposal, or sent for reprocessing.

Reprocessing of the fuel involves extracting the specific elements of interest. This has the benefits of not only recycling the fuel, but it also can reduce the volume and radioactivity of the waste material that must ultimately be sent for permanent disposal.

The most common method for reprocessing fuel involves the dissolving of the spent fuel in nitric acid and mixing with tributyl phosphate based oil. The oil forms compounds with uranium and plutonium bringing them into the oil. The oil and acid can then be separated. As with the other uranium enrichment and fuel fabrication plants the levels of discharges vary depending on throughput.

2.2.2 Nuclear defence establishments

Second to nuclear reactors, one of the most common known sources of radioactive discharges is associated with nuclear defence. This not only includes nuclear weapons, but also nuclear powered submarines.

Nuclear weapons derive their destructive force from either fission, e.g. the fission bombs used at Hiroshima and Nagasaki, or a combination of fission and fusion (the process where two light nuclei combine together releasing vast amounts of energy), or a thermonuclear bomb as it is more commonly known.

Historically nuclear weapons tests used to constitute a large source of radioactive discharges to the atmosphere, however, following the signing of the Comprehensive Nuclear-Test-Ban Treaty in 1996 this has reduced significantly.

In the present day, the majority of the nuclear weapon related discharges are associated with the maintenance of the country's nuclear deterrents. It should be noted that in some countries weapons testing still takes place, however, this is only a small portion of the world's nuclear capable countries. As an example table 2.2 shows a summary of the typical discharges associated with the maintenance of the UK's nuclear deterrent [7, 8].

The majority of nuclear powered submarines contain pressurised water reactors, these are of a similar design to the commercial civil reactors discussed in section 2.1,

Table 2.2. Typical gaseous radioactive discharges from the UK atomic weapons establishment.

Radionuclide	Discharge (MBq yr^{-1})
Tritium (^{3}H)	7×10^{5}
Alpha	0.04
Particulate beta	0.02

however, of a much smaller scale. The use of nuclear reactors allows the submarine to stay submerged for entire missions lasting up to several months.

Unlike civil reactors, nuclear powered submarines are unable to discharge radioactive particulate to the air, due to the requirement to stay submerged, and as such the majority of the discharges are aquatic in nature and directly into the sea [9, 10].

The majority of the gaseous radioactive discharges associated with nuclear powered submarines therefore occurs as a result of maintenance and decommissioning of the submarines.

Table 2.3 shows a summary of the typical discharges associated with the maintenance of the UK's nuclear powered submarines [7, 8].

2.2.3 Radiopharmaceutical production

In addition to the civil nuclear and defence industries, ionising radiation also has a role to play within the medical sector, both for diagnostic and treatment purposes. The ionising radiation originates from one of two sources:

- **Machine generated sources**. This includes x-ray and CT scanners along with external radiotherapy machines. These involve the use of a machine to both generate and carefully aim beams of radiation to produce images of the inside of the body or target a cancer.
- **Radioactive substances**. This includes the injection, ingestion or implanting of a radioisotope into the human body for either diagnostic purposes or therapy. These isotopes are called radiopharmaceuticals.

There are no radioactive discharges associated with the production of machine generated sources, however the production of radiopharmaceuticals can result in a range of radioactive discharges.

Radiopharmaceuticals can be manufactured by several different techniques depending on the specific radioisotope being produced [11]. These include:

- **Nuclear reactors**. As discussed in section 2.1, radioisotopes are produced in the form of fission fragments as a by-product of the nuclear fission process. This includes iodine-131 (^{131}I), molybdenum-99 (^{99}Mo) and xenon-133 (^{133}Xe). When producing radionuclides for medical purposes using nuclear fission, the process must be carefully controlled to minimise impurities.

Table 2.3. Typical gaseous radioactive discharges from the maintenance of the UK's nuclear powered submarines.

Radionuclide	Discharge (MBq yr^{-1})
Tritium (^{3}H)	500
Carbon-14 (^{14}C)	910
Argon-41 (^{41}Ar)	6.73
Other beta/gamma	0.02

In addition to fission fragments, radioisotopes can be produced in a nuclear reactor by bombarding a target material with neutrons. This can result in a nuclear reaction within the target material converting it into a radioisotope, an example of which is presented below. The nuclear reaction will be influenced by the energy of the neutron, and the composition of the target material.

$$n + {}^{A}_{Z}X \rightarrow {}^{A+1}_{Z}X$$

- **Particle accelerators**. Radioisotopes may also be produced by bombarding a target material with high-energy charged particles in a particle accelerator, an example of which is presented below. In such cases the yield of radioisotopes tends to be small in comparison to those produced within a nuclear reactor.

$$p + {}^{A}_{Z}X \rightarrow {}^{A+1}_{Z+1}Y$$

- **Radionuclide generator**. Radioisotopes with a short half-life are produced using a radionuclide generator which separates the progeny radionuclide from its parent by chemical or physical separation. Due to the short half-life of the progeny this process must be undertaken in a short period prior to the administering of the radioisotope to the patient. An example includes technetium-99m (^{99m}Tc) and an illustration of the radionuclide generator is provided in figure 2.5.

Gaseous radioactive discharges arise from the ventilation of radioactive or potentially radioactive areas and process equipment from the above technologies, with the levels and radioisotope breakdown of discharges varying depending on throughput and type of radiopharmaceutical being manufactured.

Table 2.4 shows a summary of the typical discharges associated with the production of radiopharmaceuticals at GE Healthcare's Amersham Site in the UK [7, 8]. The Amersham Site is GE Healthcare's main establishment in the UK and consists of a range of plants for manufacturing radioisotopes for use in diagnostic imaging in medicine and research.

2.2.4 Waste treatment facilities and landfill sites

Along with the generation of gaseous radioactive discharges, the majority of the previous examples also produce solid radioactive wastes. These wastes are either sent directly to disposal, or may require treatment prior to disposal. The disposal route for these solid wastes is dependent on several factors including, in particular, the radioactivity of the waste.

Within the UK, radioactive waste is classified under the broad categories identified in figure 2.6. In addition to those specified there are also 'exemption provisions', which allow wastes that are radioactive to be exempt from several requirements under the UK regulations for disposal of radioactive waste.

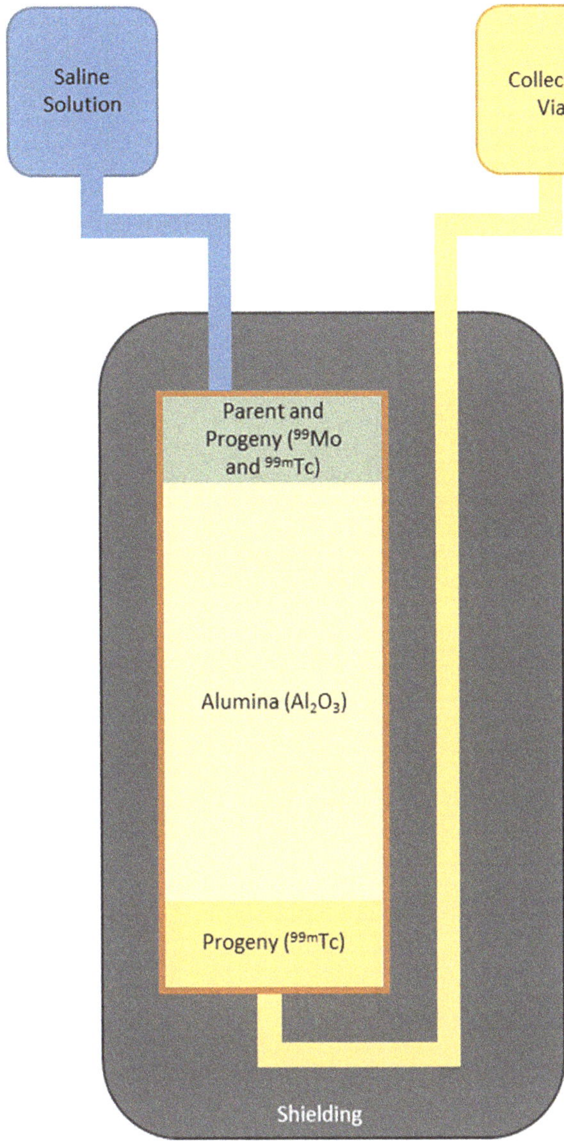

Figure 2.5. ^{99m}Tc radionuclide generator.

For instance, limited amounts of solid radioactive waste can be disposed of conveniently and without causing environmental harm if it is mixed with large quantities of non-radioactive waste. This was historically called very low level waste or 'dustbin disposal'.

The current UK strategy for the management of radioactive waste is:

- Higher activity wastes (HLW, ILW, LLW with no current disposal route, and ultimately spent fuel)—interim storage till a geological disposal facility (GDF) becomes available.

Table 2.4. Typical gaseous radioactive discharges from the production of radiopharmaceuticals at GE Healthcare's Amersham Site in the UK.

Radionuclide	Discharge (MBq yr^{-1})
Alpha	0.08
Short lived radionuclides (half-life < 2 h)	5970
Tritium (^{3}H)	1.08
Radon (^{222}Rn)	2.37×10^{6}
Other radionuclides	5.83

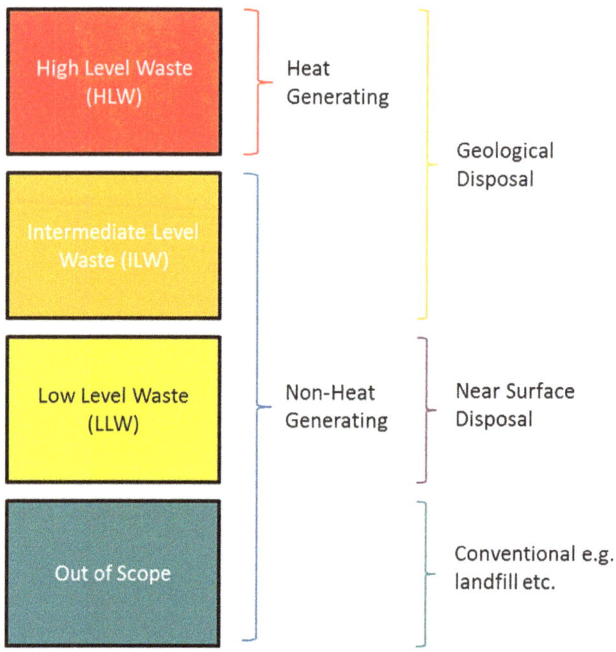

Figure 2.6. Disposal routes for radioactive waste in the UK.

- Lower activity wastes (LLW and exempt wastes)—disposal via a near surface disposal facility or permitted landfills (if certain criteria are met).

As mentioned earlier, prior to disposal the wastes may require treatment, this may be to either segregate or reduce the final waste quantities for disposal, or to immobilise the waste should it have the potential to cause contamination.

The processes used for the treatment of radioactive wastes, and those associated with its final disposal, such as near surface disposal may result in the generation of gaseous radioactive discharges [12]. Examples of the typical levels of gaseous radioactive discharges are presented in table 2.5 [7, 8].

Table 2.5. Typical gaseous radioactive discharges from the UK's metals recycling facility.

Radionuclide	Discharge (MBq yr^{-1})
Alpha (particulate)	4.30×10^{-3}
Beta (particulate)	1.49×10^{-2}

Table 2.6. Typical gaseous radioactive discharges from the UK's low level waste repository (near surface disposal facility).

Radionuclide	Discharge (MBq yr^{-1})
Alpha	8.33×10^{-3}
Beta	6.15×10^{-2}

The metals recycling facility is an example of waste treatment, where the metallic radioactive waste is sorted, segregated, monitored and size reduced, followed by surface decontamination. The intent of the process is to return the decontaminated metals to the open market for recycling, whilst the by-products of the decontamination process (secondary wastes) are disposed of as either LLW at a near surface repository, or VLLW at a landfill. Small amounts of gaseous radioactive discharges are also generated due to the decontamination process, as seen in table 2.6.

Those radioactive discharges associated with the UK's near surface disposal facility are related to the encapsulation of wastes prior to disposal and retrieval and repackaging operations of legacy wastes into more suitable containers.

2.2.5 Research establishments

In addition to the commercial or defence related operations, several countries have active nuclear research programmes. The type of research can vary significantly with some typical examples provided below:

- **New fuel cycle technology**. This includes research into new technologies for fuel enrichment or reprocessing.
- **New nuclear reactor technology**. This can involve the development of smaller modular reactors, reactors that use new fuel types, fast breeder reactors, or simply improvements on existing designs.
- **Waste treatment and disposal related research**. One of the main areas of research is into new technologies for the treatment of radioactive wastes and final disposal. Of interest are those radioactive wastes with no existing disposal route (orphan or problematic wastes).
- **Decommissioning and spent fuel**. Along with developing new reactor technologies, a large amount of research also goes into finding solutions to deal with the existing nuclear legacy. This includes technologies to assist in the decommissioning of the existing nuclear facilities/plants and management of the spent fuel.

- **New energy sources**. Some of the largest international nuclear research collaborations are in relation to nuclear fusion research, with the main goal of creating a commercially viable fusion reactor. Two of the largest projects include the International Thermonuclear Experimental Reactor (ITER) in France and the National Ignition Facility in the USA.

Depending on the nature of the research these programmes may result in the production of gaseous radioactive discharges, although these will vary based on the technology and will be far less than those associated with the commercial operating reactors and defence operations.

2.3 Discharges from non-nuclear facilities

In addition to the discharges associated with the nuclear industry, there are several additional sources of radioactive discharges from outside the nuclear industry. These tend to be minor contributors to gaseous radioactive discharges, and vary from facility to facility based on throughput and activities.

- **Education establishments**. Numerous universities undertake teaching and research activities involving the use of radioactive materials. Because of these activities there may be minor gaseous and aquatic radioactive discharges, however, these will be relatively minor.
- **Hospitals**. As mentioned in section 2.2.3 ionising radiation has a role to play within the medical sector, both for diagnostic and treatment purposes. The use of radiopharmaceuticals in hospitals may result in the production of minor quantities of gaseous and aquatic discharges.
- **Technologically enhanced NORM industries**. NORM is an acronym for Naturally Occurring Radioactive Material. This includes all radioisotopes found naturally in the environment. Technologically enhanced NORM (TENORM), refers to NORM where the amount of radioactivity has been increased or concentrated because of industrial processes [13]. Two of the most common sources of TENORM stem from the production of oil and gas and manufacture of phosphate based fertilisers.

  In the case of the former, TENORM becomes deposited in varying concentrations in different parts of the process, including pipe-line scrapings as well as sludge accumulating in tank bottoms, separators, tanks and waste pits. Radium-226 and radium-228 are the nuclides of primary concern in oil production. These decay through various ‘progeny’ radionuclides, before becoming stable isotopes of lead. Radon-222 is also present in varying concentrations and can appear dissolved in hydrocarbon and aqueous phases or can be found in gaseous streams. Other radionuclides such as lead-210 and polonium-210 can also be found. Workers in the oil and gas industry routinely work with TENORM and have to perform decontamination work to remove the build-up of scales in pipework.

  The majority of the TENORM produced as a by-product of these industrial processes is disposed of as solid or liquid waste, however, there may also be some gaseous discharges. Next to the nuclear industries TENORM discharges are one of the largest sources of radioactive discharges [14].

2.4 Physiochemical form and duration of radioactive discharges

Gaseous radioactive discharges are broadly divided into two different physical categories:

- **Aerosols**. This is a suspension of solid or liquid particles or particulates in a gas such as air, the most common example of which is dust. The majority of the radionuclides discharged into the atmosphere are in this form and include vapours such as tritiated water vapours, along with solid particulates such as radioisotopes of iodine or cobalt.
- **Gases**. Along with aerosols, several of the radioactive discharges are in the form of pure gases. These include radioactive isotopes of nitrogen, oxygen and carbon dioxide, along with noble gases such as krypton and xenon.

The physical form of the discharge has an impact on how the discharge disperses in the environment. Whilst both gases and aerosols move by the process of diffusion, aerosols undergo additional processes such as deposition. These are explored further in chapter 3.

The chemical properties of the discharge can also effect how the discharge disperses in the environment due to chemical reactions such as tritium gas (^{3}H) combining with oxygen to form tritiated water vapour (an aerosol). In addition, the chemical properties influence how they interact with the environment, humans, plants and animals. This is explored further in chapters 4 and 5.

The dispersion of the gaseous radioactive discharges and subsequent interaction with the environment is also influenced by the duration of the release. In general, these are divided into two categories:

- **Continuous releases**. These are planned discharges of radioactive aerosols and gases into the environment. As the radioactive discharges are generated, for instance from a nuclear reactor, they are usually filtered to remove the majority of the aerosolised particulate, prior to being discharged. The discharges may also be passed through additional treatment systems, such as delay beds, to allow radioactive gases such as krypton-85 (^{85}Kr) to decay for a period prior to being released into the environment. As these radioactive discharges are continually produced, they are continuously discharged from the reactor, whilst they are operating. This can result in a build-up of radioactive materials in the environment, however, this tends to be small.
- **Short term releases**. Short term releases may be both planned or due to an accident, such as in the case of Fukushima or Chernobyl. In these cases, the discharge is usually only for a short period < 24 h. This short duration influences the dispersion in the environment, and although the initial increase in radionuclide concentration in the environment may be high, any build-up tends to decrease as a function of time.

2.5 Source term

When assessing the impacts to the human environment from gaseous radioactive discharges, the first step is to determine or derive the ‘source term’. This is broadly defined as the amount of radioactive material released into the environment and is presented as a breakdown of the radioactivity (in Bq) by nuclide, normally along with details of the physiochemical form of the discharge and release duration.

The source term can either be modelled or derived from monitoring/sampling data. Modelled data tends to be used when undertaking some form of prospective assessment of the radioactive discharge, namely prior to the discharge occurring, whilst retrospective assessments occur after the discharge occurs and is based on measured data from the discharge.

Both methods of determining the source term have their positives and negatives. Modelled source terms consider an isolated discharge, namely from a single source or point, with little consideration of accumulation of radioactive material in the environment from an existing source of radioactivity. They also are subject to errors due to the assumptions made in the model.

Source terms determined by direct measurement can consider an isolated discharge or provide a holistic view of the accumulation of radioactivity in the environment, depending on what is sampled/measured. However, this is also subject to error due to sampling and analytical variability.

2.6 Summary

- Background radiation refers to the annual exposure to a member of the public from a variety of sources of ionising radiation. The average worldwide value of background radiation is 2.4 mSv yr^{-1}, where an mSv is a unit of radiation exposure.
- Gaseous radioactive discharges make up only a very small portion (~0.01%) of the exposures. They originate from several different sources, the most common of which are in relation to the nuclear industry.
- A nuclear reactor generates heat via a process called nuclear fission. This is a reaction where the nucleus of a heavy fissile nuclei, is split into two approximately equal parts, called fission fragments, releasing large quantities of energy. The fission fragments are unstable releasing neutrons, which go on to interact with other fissile nuclei, causing further fission, resulting in what is known as a chain reaction.
- A nuclear reactor is made up of the following key components:
  - fuel
  - moderator
  - control rods
  - coolant
  - turbine.

  Depending on the design they may also have a steam generator and condenser.

- In most facilities handling radioactive materials, the gaseous discharges arise from the ventilation of radioactive or potentially radioactive areas and process equipment.
- The uranium fuel cycles consist of the following stages:
  - mining of natural uranium
  - uranium enrichment
  - fuel fabrication
  - fuel used in a nuclear reactor
  - storage of spent fuel
  - reprocessing of spent fuel.

  At each of these stages gaseous radioactive discharges are generated with the exact levels of discharges varying depending on the throughput.
- Nuclear weapons derive their destructive force from either fission, (fission bombs), or a combination of fission and fusion (thermonuclear bombs).
- The majority of the nuclear weapon related discharges are associated with the maintenance of a country's nuclear deterrents. However, a small number of countries still undertake weapons testing.
- Gaseous radioactive discharges associated with nuclear powered submarines occur because of maintenance and decommissioning of the submarines. Due to the requirement to stay submerged during operation, the majority of the discharges are aquatic in nature and directly into the sea.
- Radiopharmaceuticals include radioisotopes injected, ingested or implanted into the human body for both diagnostics and therapy.
- There are three main methods for producing radiopharmaceuticals:
  - nuclear reactors;
  - particle accelerators;
  - radionuclide generators.

  The levels and radioisotope breakdown of the gaseous discharges associated with these methods vary depending on throughput and type of radiopharmaceutical being manufactured.
- The treatment and disposal of solid radioactive wastes may result in the generation of gaseous radioactive discharges.
- Several countries have active nuclear research programmes; these produce a small portion of the gaseous radioactive discharges. The key research areas relate to the development of new reactor technologies, solutions to the nuclear legacy and new energy sources.
- Along with the discharges from the nuclear industry there are several additional sources of radioactive discharges. These include:
  - education establishments;
  - hospitals;
  - TENORM industries (including oil and gas and fertiliser production).
- Gaseous radioactive discharges can be made up of aerosols (particulate matter in air) and gases. The physiochemical form of the discharge influences how it disperses in the environment along with how it interacts with the environment, humans, plants and animals.

- Discharges can either be planned or unplanned (accidental) and may be continuous or short duration releases. This has an impact on the dispersion and build-up in the environment.
- The amount of radioactive material released into the environment is called the 'source term' and is presented as a breakdown of the radioactivity (in Bq) by nuclide, normally along with details of the physiochemical form of the discharge and release duration.
- The 'source term' can either be modelled or derived from direct sampling or monitoring of an existing facility. This is dependent on the availability of data and type of assessment being undertaken. Both modelled and measured source terms can be subject to errors.

References

[1] Public Health England 2016 *Ionising Radiation Exposure of the UK Population 2010 Review*

[2] Darwish A, Karunakara N and Mustapha A O 2013 Teaching about natural background radiation *Phys. Educ.* **48** 506–11

[3] Martin A, Harbison S, Beach K and Cole P 2012 *An Introduction to Radiation Protection* 6th edn (Boca Raton, FL: CRC Press)

[4] World Nuclear Association *Environmental Aspects of Uranium Mining* http://world-nuclear.org/information-library/nuclear-fuel-cycle/mining-of-uranium/environmental-aspects-of-uranium-mining.aspx [accessed 9 April 2017]

[5] Bryant P 2014 Chemical toxicity and radiological health detriment associated with the inhalation of various enrichments of uranium *J. Radiol. Prot.* **34** N1–6

[6] The nuclear fuel cycle *Encyclopaedia Britannica* https://britannica.com/technology/nuclear-reactor/Three-Mile-Island-and-Chernobyl#toc302441 [accessed 9 April 2017]

[7] Food Standards Agency 2016 *Radioactivity in Food and the Environment* RIFE-21

[8] Food Standards Agency 2011 *Radioactivity in Food and the Environment* RIFE-17

[9] National Academy of Sciences 1971 *Radioactivity in the Marine Environment* 1st edn (Washington, DC: The National Academy of Sciences)

[10] Curren T 1988 *Nuclear Powered Submarines: Potential Environmental Effects.*

[11] World Health Organisation 2008 *Radiopharmaceuticals Final text for addition to The International Pharmacopoeia* QAS/08.262/FINAL

[12] Low Level Waste Repository 2009 *LLWR Environmental Safety Case—LLWR Authorisation Schedule 9, Requirement 3 Study,* NNL(09)10297 Issue 1

[13] World Nuclear Association *Naturally-Occurring Radioactive Materials (NORM)* http://world-nuclear.org/information-library/safety-and-security/radiation-and-health/naturally-occurring-radioactive-materials-norm.aspx [accessed 30 April 2017]

[14] OSPAR Commission 2015 *Discharge of Radionuclides from the Non-nuclear Sectors in 2013.*

Chapter 3

Aerosol physics and dispersion modelling

As discussed in chapter 2 the main sources of gaseous radioactive discharges arise from the ventilation of radioactive or potentially radioactive areas and process equipment within facilities handling radioactive materials. These discharges normally take place at height from a stack, to maximise dispersion into the environment.

The airborne discharges usually consist of a combination of aerosols and gases [1]. In both cases these move by the process of diffusion, however, aerosols undergo additional processes such as deposition.

The processes by which aerosols move within the physical world are explored within this chapter.

3.1 Aerosols—particulate size and particulate concentration

Aerosols are broadly defined as a suspension of particulate matter in air, where particulate matter can be a solid or liquid or a combination of them both. They are formed by the conversion of a gas to particulate, by the disintegration of liquids or solids, or by a resuspension of powder.

Aerosols are normally defined by two parameters, particulate size and particulate concentration. These are explored further in the subsequent sections.

3.1.1 Particulate size

The size of the aerosol particulate can have an impact on how the aerosol interacts with the surrounding gas (such as air).

The diameter of the particulate can range from molecular clusters of the order of 1 nm to fog droplets and dust particles which can be as large as 100 μm.

Aerosol particulates can be spherical in shape due to their growth by condensation in the liquid phase such as tritium vapour. However, it is more common in the case of radioactive discharges that they are irregular in shape, making it difficult to define the size and shape of the particulate.

doi:10.1088/2053-2563/aafa6dch3

It is therefore more convenient to express the particulate size in terms of particulate volume.

As mentioned earlier, the size of the particulate affects how the aerosol interacts with the surrounding gas, this includes the transfer of heat and momentum. The transfer is determined by the Knudsen number (K_n) which is calculated as follows:

$$K_n = \frac{2l}{d_p} \tag{3.1}$$

Where l is the number of mean free paths of the gas molecules (in m), or average distance travelled by a gas molecule between interactions, and d_p is the particulate diameter.

The mean free path is calculated using the following expression from kinetic theory:

$$l = v\left(\frac{\pi m}{2kT}\right)^{1/2} \tag{3.2}$$

Where v is the kinematic viscosity of the gas (in $m^2\ s^{-1}$), m is the molecular mass of the gas (kg), k is the Boltzmann's constant (1.38×10^{-23} J K^{-1}) and T is absolute temperature (K).

For air, at ambient temperature and pressure, the mean free path l is approximately 0.065 μm.

In the case that particulate diameter is much less than the mean free path of the gas molecules ($d_p \ll 2l$), the value of K_n is large, and the heat and momentum transfer between the gas and the aerosol particulate can be modelled using molecular collision theory.

Whereas, if the mean free path of the gas molecules is much less than the particulate diameter ($d_p \gg 2l$), the value of K_n is small, and the heat and momentum can be modeled as a continuum allowing classical approaches to be used.

3.1.2 Particulate concentration (number and mass distribution)

The particulate concentration can be expressed on a number or mass basis as defined below:

- **Number basis**. The particulate number concentration is expressed as the number of particulates per unit volume of gas.
- **Mass basis**. The particulate mass concentration is expressed as the mass of particulate per unit volume of gas.

However, aerosols are usually poly-disperse, namely there is a distribution of particulates of different sizes.

For a given point in space and time we can consider the number of particulates per unit volume (dN) in the size range dp to $dp + d(dp)$ as:

$$dN = nd(dp) \tag{3.3}$$

where n is the particulate size distribution function or number of particulates per unit size range per unit volume with units of m^{-4}.

From equation (3.3) the cumulative number of particulates N in a population from zero size to some size dp can be given by:

$$N = \int_0^{dp} nd(dp) \tag{3.4}$$

This is summarised in figure 3.1.

3.2 Aerosol stability

Up until this point we have considered particulate size distribution to be stable, however, in reality this varies with time and position according to the mechanisms discussed below and highlighted in figure 3.2.

- **Formation and growth**. As a result of cooling or chemical reactions, gas phase molecules can be converted into the aerosol phase. This can either be in the form of many very small, new particulates of less than 10 nm diameter, or by condensation on existing particulates.
- **Evaporation**. Aerosol particulates may evaporate due to changes in humidity or temperature.
- **Diffusion**. This is where particulates move away from an area of high particulate concentration to an area of low particulate concentration. The smaller the size of the particulate the more rapidly they diffuse. This is as a result of Brownian motion, which is discussed further later in this chapter.
- **Coagulation**. This occurs when particulates collide and stick together as a result of Brownian motion. This is more rapid near the source of the aerosol as the concentration of particulates is highest. The rate of coagulation is proportional to the product of the concentrations of colliding particulate.
- **Sedimentation (or deposition)**. This is more significant the larger the particulate size (> 1 μm) and is a result of gravity acting on the particulate.

Although all the above mechanisms are of equal importance, two of the most relevant to the interaction with the human environment are diffusion and sedimentation [2]. These mechanisms are explored in further detail later in this chapter.

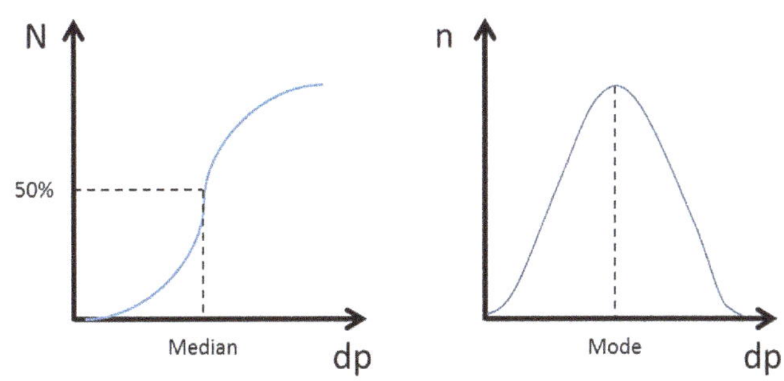

Figure 3.1. Cumulative and particulate size distribution for a population from zero size to some size *dp*.

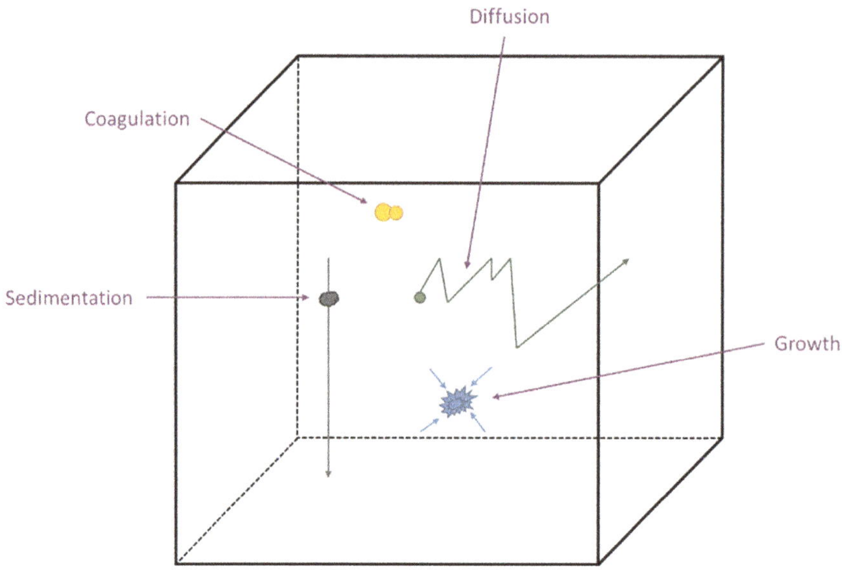

Figure 3.2. Mechanisms that impact particulate size distribution of aerosols.

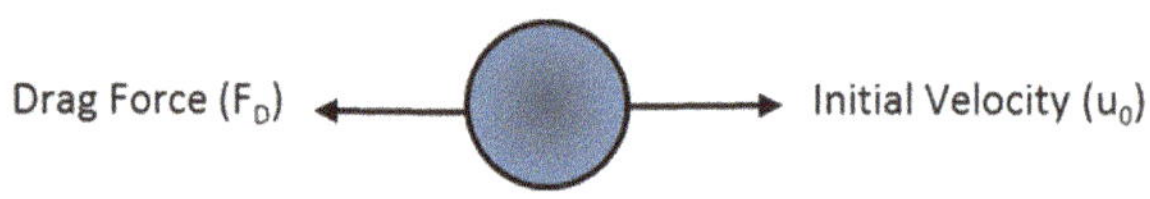

Figure 3.3. Force diagram for an aerosol particulate without external forces (in the horizontal plane).

3.3 Aerosol motion without external forces

As mentioned earlier, radioactive discharges normally take place at height from a stack. When they emerge from the stack the aerosol particulates are acted upon by a number of external forces, such as sedimentation affecting the vertical axis and the force from the ejection of the stack affecting the horizontal direction.

As a starting point let us simply consider the horizontal motion, namely the motion in a straight line, with the particulate moving into stationary air with a finite velocity u_0 (figure 3.3).

From Newton's second law we can write a force balance on the particulate as follows:

$$m_P \frac{du}{dt} = -F_D \tag{3.5}$$

where m_P is the mass of the particulate and (du/dt) is the change in velocity per unit time, or acceleration.

In classical physics the drag force can be represented by Stokes' law as:

$$F_D = 3\pi\mu d_p u \tag{3.6}$$

where μ is the dynamic viscosity, and d_p is the particulate diameter.

Therefore, substituting into the force balance (3.5) gives you:

$$m_P \frac{du}{dt} = -3\pi\mu d_p u \tag{3.7}$$

The force balance can be solved to give the particulate velocity in terms of either the distance travelled, x or the time travelled, t.

3.3.1 Velocity as a function of distance

Using equation (3.7) we can substitute for x as follows:

$$m_P \frac{du}{dx} \cdot \frac{dx}{dt} = -3\pi\mu d_p u \tag{3.8}$$

However as:

$$u = \frac{dx}{dt} \tag{3.9}$$

this can be simplified to:

$$m_P \frac{du}{dx} = -3\pi\mu d_p \tag{3.10}$$

Separating by variables and integrating gives:

$$\int_{u_0}^{u} du = \frac{-3\pi\mu d_p}{m_p} \int_0^x dx \tag{3.11}$$

$$u - u_0 = \frac{-3\pi\mu d_p x}{m_p} \tag{3.12}$$

Now as the mass, m_p of the particulate is given by:

$$m_p = \rho_p \cdot \frac{\pi d_p{}^3}{6} \tag{3.13}$$

we can simplify expression (3.12) to:

$$u - u_0 = -\frac{x}{\tau} \tag{3.14}$$

where T is the relaxation time or:

$$\tau = \frac{\rho_p d_p{}^2}{18\mu} \tag{3.15}$$

3.3.2 Velocity as a function of time travelled

Similarly, using equation (3.7) we can rearrange the force balance and integrate as follows:

$$\int_{u_0}^{u} \frac{du}{u} = \frac{-3\pi\mu d_p}{m_p} \int_{0}^{t} dt \tag{3.16}$$

Substituting for m_p from equation (3.13) and solving gives:

$$ln\left(\frac{u}{u_0}\right) = -\frac{18\mu t}{\rho_p d_p^2} \tag{3.17}$$

$$u = u_0 . \exp\left(-\frac{t}{\tau}\right) \tag{3.18}$$

Equations (3.15) and (3.18) both provide a useful of means of modelling the velocity of the aerosol particulate. However, they are based on a number of simplifications and assumptions which leads to inaccuracies. These include:

- Mechanisms of diffusion, and sedimentation will be active. Also, electrostatic charges may exert force on particulates that modifies motion.
- Evaporation reducing the particle size and therefore the relaxation time, T will decrease with distance.
- High velocities—at higher velocities Stokes' law breaks down and loses its validity. Stokes' law also only applies for particulate with a low K_n (i.e. $d_p \gg 2l$).
- Entrainment of air, where air is being forced along with the aerosol particulates reducing the drag force acting on them.
- Divergence of flow. The calculation assumes all particulates move in the same direction when in reality they do not.
- Coalescence of the particulates may occur due to air turbulence.
- We do not have uniform sized particulates, but rather a distribution of sizes.

3.4 Sedimentation

We have previously looked solely at the horizontal motion of the aerosol particulate, negating any external forces. However, as mentioned, modelling the motion in such a way can present several inaccuracies.

Sedimentation is one of the key parameters affecting the aerosol stability, and affects the vertical motion of the particulate. It is relevant for larger particulates (> 1 μm) and is a result of gravity.

In a similar fashion to section 3.4, consider the motion of the particulate in the vertical direction only (figure 3.4).

From Newton's second law we can write a force balance on the particulate as follows:

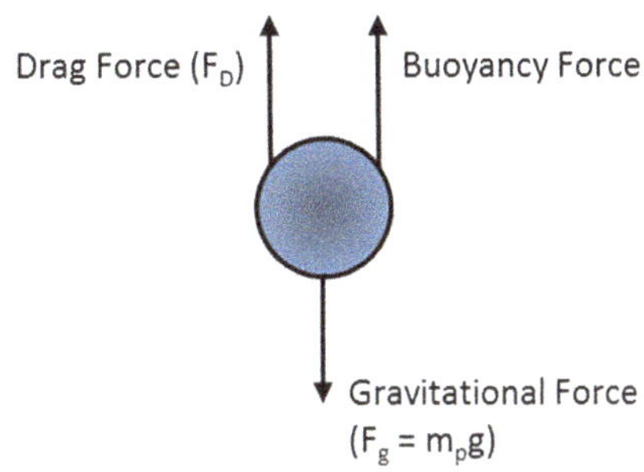

Figure 3.4. Force diagram for an aerosol particulate (in the vertical plane).

$$m_P \frac{dv}{dt} = m_p g - F_B - F_D \tag{3.19}$$

where u is the velocity in the vertical direction and g, is the acceleration due to gravity (9.81 m s^{-2}).

However, from classical physics we know that buoyancy force can be expressed as:

$$F_B = \frac{\pi d_p{}^3}{6} . \, g . \, \rho_f \tag{3.20}$$

where ρ_f is the fluid density.

Upon release, the aerosol particulate will accelerate with an increasing velocity u. As this velocity increases so does the value of F_D, until at some point:

$$m_P \frac{dv}{dt} = 0 \tag{3.21}$$

At this time the particulate has reached what is known as its terminal settling velocity, u_T. As such:

$$F_D = m_p g - F_B \tag{3.22}$$

Substituting equations (3.6), (3.13) and (3.20) into equation (3.22) gives:

$$3\pi\mu d_p u_T = \frac{\pi d_p{}^3}{6}(\rho_p - \rho_f)g \tag{3.23}$$

Therefore, the terminal settling velocity, u_T is equal to:

$$u_T = \frac{d_p{}^2(\rho_p - \rho_f)g}{18\mu} \tag{3.24}$$

This expression is useful for estimating the time it takes for the particulate to reach the ground from release of a stack along with the distance that the particulate can travel in the horizontal direction prior to reaching the ground.

It should be noted that this equation only applies for particulates with a low K_n i.e. $d_p \gg 2l$. For smaller particulates, a correction factor, known as the Cunningham correction factor should be applied.

Due to the application of Stokes' law, this expression is also only valid for particulates with a Reynolds number ($R_e < 1.0$) or low velocity, where:

$$R_e = \frac{\rho_p u d_p}{\mu} \tag{3.25}$$

3.5 Diffusion and the Gaussian plume

Along with sedimentation, one of the key parameters affecting aerosol stability which is particularly relevant to the interaction with the human environment is diffusion.

This mechanism helps the particulate travel over large distances, especially after the force from the initial ejection of the stack has diminished due to the drag force.

The movement of particulate is primarily attributed to 'Brownian motion', which is the random movement of small particulates due to collisions with gas molecules [3]. This causes a net migration of particulates from regions of higher concentration to regions of lower concentration, in a similar way to if you added a drop of red food colouring to a glass of water, you would slowly see the colour disperse from the point where it was initially dropped till it fills the whole glass.

The local net flux of particulate (in one dimension) can be described by Flick's law, as seen in equation (3.26) below:

$$J = -D_B \frac{dc}{dx} \tag{3.26}$$

where J is flux or number of particulates moving per unit time per unit cross sectional area, D_B is the diffusion coefficient, c is the local concentration and x is distance.

For small particulates that obey Stokes' law, the diffusion coefficient can be calculated using the Stokes' Einstein equation:

$$D_B = \frac{kT}{F_D} = \frac{kT}{3\pi\mu d_p} \tag{3.27}$$

where k is the Boltzmann's constant, T is the temperature, F_D is the drag force, μ is the dynamic viscosity, and d_p is the particulate diameter.

As mentioned earlier, Brownian motion involves the migration of particulates from regions of higher concentration to regions of lower concentration. This can be expressed mathematically as:

$$\frac{dc}{dt} = -\frac{dJ}{dx} \tag{3.28}$$

Substituting in Flick's law gives:

$$\frac{dc}{dt} = -\frac{d}{dx}\left(-D_B\frac{dc}{dx}\right) \tag{3.29}$$

$$\frac{dc}{dt} = D_B\frac{d^2c}{dx^2} \tag{3.30}$$

Solving equation (3.30) for N particles released at $x = 0$ and time $t = 0$ gives the following:

$$c(x, t) = \left(\frac{N}{2\pi D_B\sqrt{t}}\right)\exp\left(\frac{-x^2}{4D_Bt}\right) \tag{3.31}$$

This expression is a Gaussian distribution for concentration along the x-axis as a function of time t. For instance, imagine we have a pipe with no flow; if we release an aerosol in the middle, then we would have the concentration profiles developing shown in figure 3.5.

This forms the foundation of what is known as the Gaussian plume model. However, this takes account of multiple dimensions (x, y and z) rather than a single dimension.

3.6 Dispersion modelling

From section 3.5, we know that following the release of a radioactive discharge from a stack, the aerosol or gas will disperse due to diffusion, however, in reality the rate of diffusion may be much greater than that due to just Brownian motion. This is a result of turbulence, which greatly enhances the transport by diffusion. For particles > 5 μm the Brownian effects are negligible, and turbulent diffusion is dominant, however, for particles < 5 μm the Brownian diffusion effects become increasingly important.

There are several models which provide a detailed representation of the physical processes of turbulent diffusion, such as the Lagrangian puff model or Eulerian grid

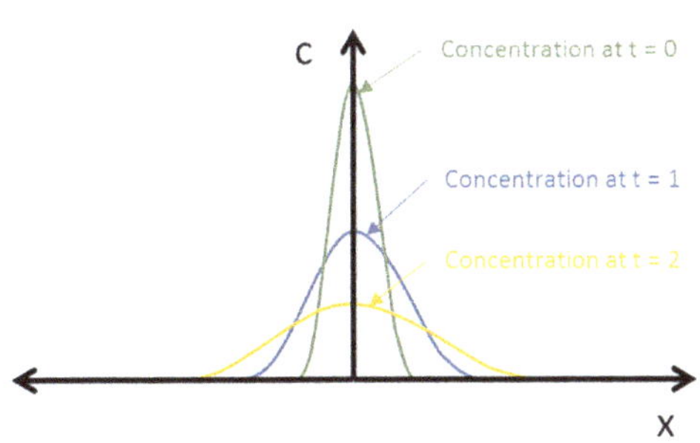

Figure 3.5. Concentration distribution at various time intervals.

model. However, these are generally computationally demanding, and depending on the application may not be required.

In the case of long duration releases, simplifying assumptions can be made, such as ignoring the direction of the wind. In these cases, the most common model used is known as the Gaussian plume model, which is widely adopted due to its simplicity, and its sufficient accuracy given the uncertainties in the parameters used.

3.6.1 Gaussian plume model

The Gaussian plume model assumes that any aerosols or gases that are discharged into the atmosphere are carried along by the wind and dispersed by turbulent diffusion [4]. As previously mentioned, one of its main simplifications is assuming the meteorological factors remain constant during the travel time.

The model describes the dispersion of material by standard deviations σ_y and σ_z in the crosswind and vertical directions, respectively. Horizontally, the dispersion process is unlimited and is affected by turbulence, which draws its energy from the influences of meteorological factors. This is summarised in figure 3.6.

This model applies to the atmospheric diffusion of a neutrally buoyant plume over a flat plane from an isolated stack. The model gives good estimates for a stack height up to and including 200 m, and for distances up to 100 km from the source in the wind and crosswind direction.

The basic equation of using a Gaussian plume model for an elevated release is as follows:

$$c(x, y, z) = \left(\frac{Q}{2\pi u_{10}\sigma_z\sigma_y}\right)\exp\left(\frac{-y}{2\sigma_y^2}\right)F(h, z, A) \qquad (3.32)$$

where c is the airborne radioactivity concentration at a point x m from the release (in Bq m^{-3}), Q is the total radioactivity of the material released (in Bq), u_{10} is the wind speed at 10 m above ground (in m s^{-1}) and, σ_z and σ_y are the standard deviations of the vertical and horizontal Gaussian distributions (in m). The function F includes the effect of reflections due to the mixing layers in the atmosphere and is a function of release height, h, and mixing layer depth A.

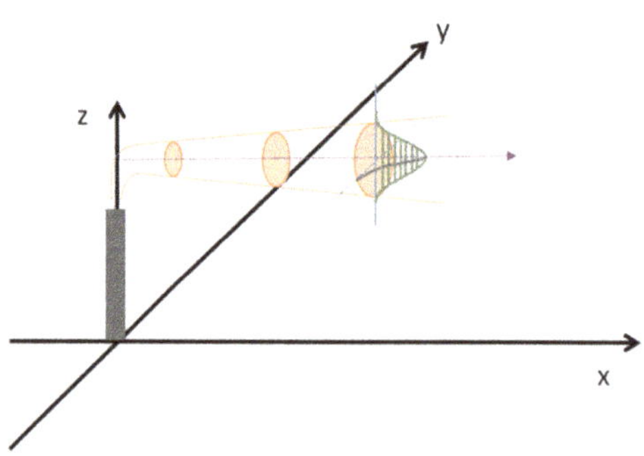

Figure 3.6. Gaussian plume model.

The wind speed u_{10} and vertical standard deviation σ_z depend on the meteorological conditions at the time of release. σ_z is also a function of downward distance and ground roughness, where ground roughness is dependent on the terrain type.

The horizontal standard deviation σ_y can be approximated as a function of wind speed and downwind distance.

The Gaussian plume model tends to be used for modelling long term continuous releases of radioactive aerosols and gases, for instance when looking at the effect of annual discharges from a nuclear power station.

It can also be used for modelling short-term releases, however, it has its limitations due its assumptions around constant meteorological conditions, stable and neutral conditions, simple terrains and its inability to model the wake affects of buildings surrounding the stack.

3.6.2 Calculating activity concentrations in air

Following the discharge of a known quantity of radioactive aerosols and gases from a stack, we need to be able to calculate the airborne concentration of radioactivity for each radionuclide at a distance from the source due to the atmospheric dispersion. This allows us to look at the potential interactions with the human environment. These interactions are called exposure pathways and are discussed in further detail in chapter 4.

There are several standard industry tools available for calculating the airborne concentration of radioactivity which use techniques such as the Gaussian plume model. A few of which are summarised below:

- **Hand calculation techniques**. When modelling simple scenarios or small releases of radioactivity it may be appropriate to use hand calculations techniques. One of the most commonly used is detailed in the UK's former National Radiological Protection Board Publication R91 [4]. This includes a set of standard dispersion coefficients, for a variety of release periods, weather conditions and stack heights. These dispersion coefficients have been calculated using a simple Gaussian plume model, and allow the user to simply multiply a known release by a coefficient to give an estimated airborne concentration of radioactivity.
- **PC-CREAM 08**. is an updated version of the Consequences of Releases to the Environment Assessment Methodology (CREAM) described in the European Commission (EC) report RP 72 [5]. The software consists of a suite of mathematical models for calculating the transfer of radionuclides through the atmospheric and aquatic environment providing an estimated activity concentration in various media following a continuous release. The airborne model is also based on a simple Gaussian plume model [2].
- **ADMS**. is an aerial dispersion software designed for environmental impact assessments and is applicable to both routine and accidental discharges [6]. It was developed in collaboration with the Met Office. The software uses an asymmetric Gaussian model for the vertical dispersion of contaminants which is more realistic under convective meteorological conditions than the

symmetric Gaussian model used in R91 and PC-CREAM. ADMS includes algorithms which take account of plume rise, building wake effects, complex terrain, and time varying and jet releases.

- **Computational fluid dynamic (CFD)**. Where the source of the radioactive discharges and human population are located within a few hundred meters from each other, and there is a particularly complicated geometry due to either building or terrain, then CFD based models may be required. These models are computationally intense and time consuming to create and as such are normally only used when a PC-CREAM or ADMS would not be suitable.

3.6.3 Far field models

So far, we have predominantly looked at the dispersion of radionuclides from the source to the nearest population, typically within 10 km, however, in reality that dispersion may happen over large distances, including crossing the boundaries of states and countries. Above distances of 20–30 km the models described in section 3.6.2, start to lose their validities due to difficulties in accounting for changes in wind direction and atmospheric stability during the passage of the plume over extended distances.

In such scenarios, simple hand calculation based models have been developed. These models are generic in nature and apply conservative assumptions to account for the wind direction changes and atmospheric stability. An example of which is described in National Radiological Protection Board Publication R123 [7]. These models tend to be applied when assessing the implications of radioactive discharges on neighbouring states or countries.

3.7 Summary

- Radioactive airborne discharges usually consist of a combination of aerosols and gases. In both cases these move by the process of diffusion, however, aerosols undergo additional processes such as deposition.
- Aerosols are broadly defined as a suspension of particulate matter in air, where particulate matter can be a solid or liquid or a combination of them both.
- Aerosols are normally defined by two parameters, particulate size and particulate concentration.
- The particulate size distribution varies with time and position according to the following mechanisms:
    - formation and growth;
    - evaporation;
    - diffusion;
    - coagulation;
    - sedimentation (or disposition).

  Two of the most relevant to the interaction with the human environment are diffusion and sedimentation.

- When radioactive discharges emerge from the stack the aerosol particulates are acted upon by a number of external forces, such as sedimentation affecting the vertical axis and the force from the ejection of the stack affecting the horizontal direction.
- Sedimentation is one of the key parameters affecting the aerosol stability, and affects the vertical motion of the particulate. It is relevant for larger particulates (> 1 μm) and is a result of gravity.
- Diffusion helps the particulate travel over large distances, especially after the force from the initial ejection of the stack has diminished due to the drag force.
- There are two types of diffusion, Brownian diffusion and turbulent diffusion. For particles > 5 μm the Brownian effects are negligible, and turbulent diffusion is dominant, however for particles < 5 μm the Brownian diffusion effects become increasingly important.
- The most common model used for calculating turbulent diffusion is the Gaussian plume model, which is widely adopted due to its simplicity.
    - The model describes the dispersion of material by standard deviations σ_y and σ_z in the crosswind and vertical directions, respectively. Horizontally the dispersion process is unlimited and is affected by turbulence, which draws its energy from the influences of meteorological factors.
    - The Gaussian plume model tends to be used for modelling long term continuous releases of radioactive aerosols and gases, for instance when looking at the effect of annual discharges from a nuclear power station.
    - It can also be used for modelling short-term releases, however, it has its limitations due its assumptions around constant meteorological conditions, stable and neutral conditions, simple terrains and its inability to model the wake effects of buildings surrounding the stack.
- To investigate the potential impacts to the human environment, the airborne concentration of radioactivity for each radionuclide at a distance from the source due to the atmospheric dispersion must be calculated. There are several standard industry tools available for calculating the airborne concentration of radioactivity. These include:
    - For simple scenarios hand calculations can be used such as those detailed in the UK's former National Radiological Protection Board Publication R91.
    - For continuous routine releases industry approved models such as PC-CREAM 08 can be used. This calculates the estimated activity concentration in both the atmospheric and aquatic environment.
    - For modelling short-term releases or complex terrains industry codes such as ADMS are well established. The software uses an asymmetric Gaussian model for the vertical dispersion of contaminants which is more realistic under convective meteorological conditions.
    - For modelling discharges at close range < 1 km and complex terrains then CFD modelling may be required. This is normally computationally and time intense.

- Far field models have been developed to look at the dispersion of radioactive discharges across country and state boundaries. An example of these is described in National Radiological Protection Board Publication R123.

References

[1] Martin A, Harbison S, Beach K and Cole P 2012 *An Introduction to Radiation Protection* 6th edn (Boca Raton, FL: CRC Press)

[2] Smith J and Simmonds J 2009 *The Methodology for Assessing the Radiological Consequences of Routine Releases of Radionuclides to the Environment Used in PC Cream-08* HPA-RPD-058

[3] Ounis H and Ahmadi G A 1990 A comparison of Brownian and turbulent diffusion *Aerosol Sci. Technol.* **13** 47–53

[4] Clarke R 1979 *A Model for Short and Medium Range Dispersion of Radionuclides Released to the Atmosphere* ISBN 0659511170

[5] Simmonds J, Lawson G and Mayall A 1995 *Methodology for assessing the radiological consequences of routine release of radionuclides to the environment European Commission* RP 72

[6] Cambridge Environmental Research Consultants 2016 *ADMS 5 Atmospheric Dispersion Modelling System User Guide* Version 5.2

[7] Jones H A 1981 *The estimation of long range dispersion and deposition of continuous releases of radionuclides to atmosphere* National Radiological Protection Board NRPB R123

Chapter 4

Exposure pathways

In chapters 2 and 3 we explored the sources of radioactive discharges and how they interact with the environment once they are discharged. As the aerosols and gases disperse there are a number of routes they can take before interacting with the human environment [1]. These routes are called 'exposure pathways'. In the following chapter, we will explore the main exposure pathways associated with airborne radioactive discharges, we will subsequently go on to explore the principles of radiation dose in chapter 5, where we will investigate the potential health impacts of these discharges.

4.1 The human environment

When looking at risks associated with pollution such as radioactive discharges it can be useful to look at the source, pathway, receptor model [2]. It starts with the source, namely where the discharge originates from, then looks at the various pathways the pollution travels through the human environment, before finally reaching a receptor, which may be a human, group of humans, or more than one group of humans.

Figure 4.1 shows the source, pathway, receptor model for airborne radioactive discharges. For simplicity, the receptors have been grouped together into 'members of the public', and as such includes infants, children and adults of both genders. Although the exposure pathways are the same for each of the receptors, the radiation dose may be different, this is explored further in chapter 5.

It should also be noted that in addition to members of the public, other receptors that can be impacted by airborne radioactive discharges include workers on the site where the discharges took place, and flora and fauna, however the exposure pathways may differ. In the case of workers, the exposure pathways tend to be the same as those highlighted in figure 4.1, however, ingestion is not normally considered as the food is not normally sourced from the area that may be impacted by the discharges. In the case of flora and fauna, the exposure pathways vary depending on the type of flora or fauna.

doi:10.1088/2053-2563/aafa6dch4

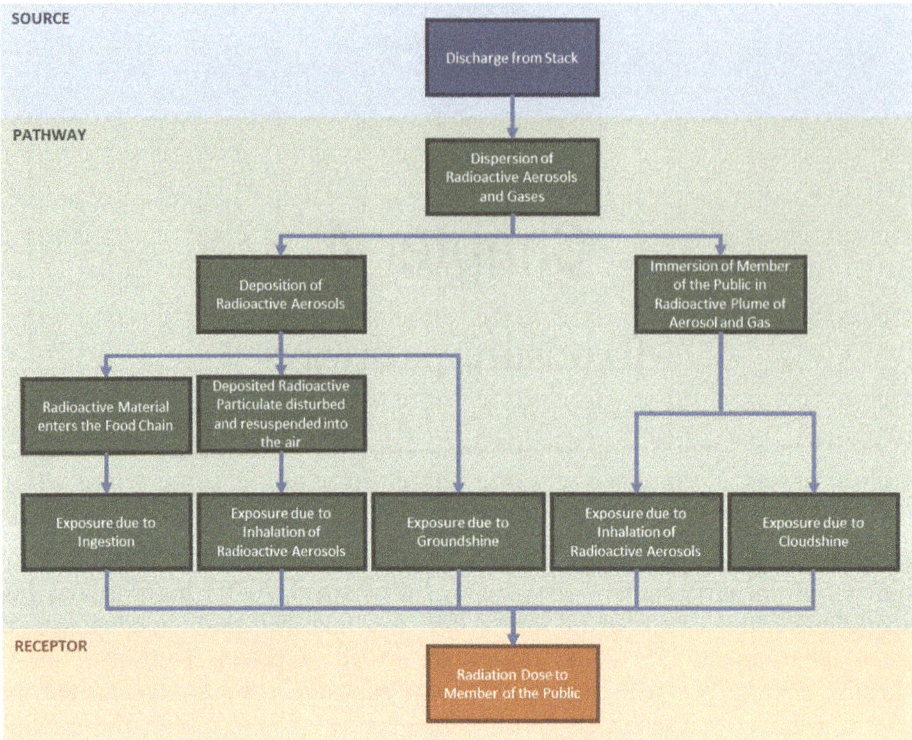

Figure 4.1. Source, pathway, receptor model for airborne radioactive discharges.

In the subsequent sections we will explore each of the exposure pathways related to the human environment further, these are summarised in figure 4.2. Those pathways related to flora and fauna will briefly be touched upon in sections 4.5 and 4.8.

4.2 Cloudshine from the plume of radioactivity

Cloudshine is the exposure an individual receives due to radiation from immersion in the 'cloud' or plume of radioactive aerosols and gases. It can also be a source of exposure even if the person is not immersed in the plume, as even if the plume is high above the individual the gamma rays can still shine down due to their ability to travel great distances.

The radiation dose from immersion is nearly always much greater than that from the cloud at a distance. As such the immersion dose is normally considered and calculated, and the dose from the cloud at a distance is normally neglected.

The radiation dose from immersion can be calculated either by measurements of the radioactivity in a plume or modelled, depending on whether a prospective or retrospective assessment of the radioactive discharges is being undertaken.

The cloudshine dose is typically calculated for gamma-emitting radioisotopes only, however it should be noted that beta and even alpha particles can also produce external doses to the skin [3].

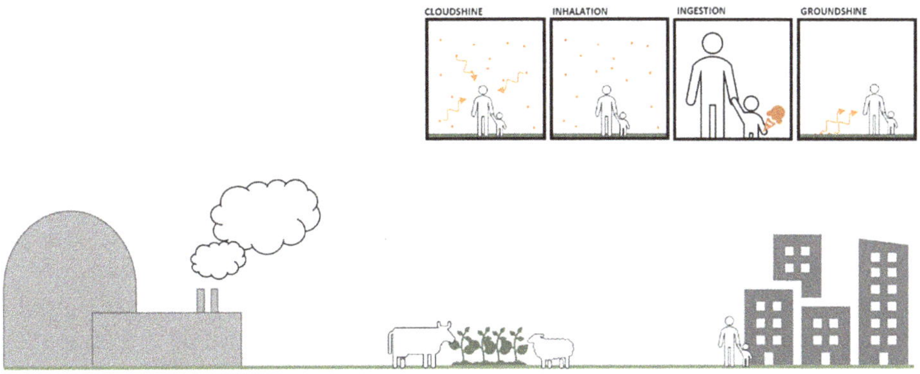

Figure 4.2. Exposure pathways for airborne radioactive discharges.

There are many ways of modelling cloudshine, such as the semi-infinite and finite cloud models [3]:

- The finite cloud model is used for lower energy gamma rays ($\leqslant$ 20 keV). It involves simulating the plume by a series of small volume sources and integrating over these sources.
- The semi-infinite cloud model is used for high-energy gamma rays (> 20 keV). This assumes that the concentration in the plume is uniform.

A number of publications have been produced based on these models which can allow indicative estimates of the cloudshine dose rate, E_c (in Sv h^{-1}), to a individual receptor to be calculated by multiplying c_i, the airborne radioactivity concentration for each radionuclide in (Bq m^{-3}), by e_i, a dose conversion coefficient (in Sv m^{-3} Bq^{-1} h^{-1}) [4, 5] and summing for all radionuclides, as seen below:

$$E_c = \sum_{i=0}^{n} c_i e_i \tag{4.1}$$

It should be noted that if the individual is indoors the structure of the building will attenuate the radiation from the plume of radioactivity outside the building. In such cases a shielding factor should be taken into account, the value of which is highly variable due to the type of building and floor level being considered.

4.3 Groundshine from the deposited particulate

As discussed in chapter 3 the process of sedimentation can result in radioactive particulates being removed from the plume and deposited on the ground. The rate of deposition can be affected by the weather conditions and physical–chemical form of the particulate. It is broadly broken down into dry and wet deposition. Deposition only applies to the aerosol particulate, and does not apply to gases which will stay within the plume [6].

4.3.1 Dry deposition

Dry deposition is the removal of the radioactive particulate from the plume purely as a result of sedimentation, as discussed in chapter 3. Namely, the deposition rate (D) is given by:

$$D = cV \tag{4.2}$$

where c is the airborne radioactivity concentration in air at ground level (Bq m^{-3}) and V is the deposition velocity (m s^{-1}) [3].

When modelling dry deposition, a single value of 10^{-3} m s^{-1} is typically used for the deposition velocity of most radioactive particulates. This value is representative of 1 μm diameter particulates. In the case of iodine, specific values for the deposition velocity are usually chosen.

4.3.2 Wet deposition

Wet deposition is the removal of radioactive particulates from the plume as a result of rainfall, this may be due to rainfall passing through the plume, which is known as washout, or removal of activity which has become incorporated into a rain cloud, which is known as rainout.

The washout rate is a function of the airborne radioactivity concentration throughout the depth of the plume being rained on. Whilst the rainout rate is influenced by condensation processes within the cloud and the diffusion rate of radioactive particulate into the rain cloud.

The wet deposition rate is usually calculated using a washout coefficient (Λ), which has been calculated to take account of both washout and rainout. Where the total amount of radioactivity in the plume Q' at time (t) subject to continuous rainfall at a constant rate is given by [3]:

$$\frac{dQ'}{dt} = -\Lambda Q' \tag{4.3}$$

In reality the interaction between the plume and rain is complex, with the assumption of continuous rain being only one of a number of possible scenarios.

4.3.3 Groundshine

The combination of wet and dry deposition leads to the build-up of radioactive particulate on the surface of the ground and soil. This has the ability to irradiate a member of the public due to gamma rays emitted by the radioactive particulate distributed in the ground and soil.

As with cloudshine the concentration of the radionuclides in the soil can either be measured or calculated, depending on whether a prospective or retrospective assessment of the radioactive discharges is being undertaken.

It is usual practice to only model the radioactive contamination which builds up in the top layers of the soil (up to 30 cm), as below this the soil will shield gamma rays to such an extent that the dose will be negligible.

Doses to the receptor due to external irradiation form beta-emitting radioactive particulates deposited on the ground and soil are sometimes considered in models. However, this tends to be only for the activity close to the surface of the soil or ground and with an energy greater than 1 MeV. This is due to the small range of the electrons in soil and also attenuation in air.

The modelling of the dose due to groundshine normally requires the use of computational software in order to simulate the physics of the radiation passing through the soil and air, along with the various scattering effects.

However, much like cloudshine, a number of publications have been produced which can allow indicative estimates of the groundshine dose, E_g, to an individual receptor to be calculated by multiplying s_i, the surface radioactivity concentration for each radionuclide in (Bq m^{-2}), by g_i, a dose conversion coefficient (in Sv m^{-2} Bq^{-1} h^{-1}) and summing for all radionuclides, as seen below [3]:

$$E_g = \sum_{i=0}^{n} s_i g_i \tag{4.4}$$

However, this only models the activity of the particulate deposited on the surface of the ground and soil and neglects those which have built up in the depths of the soil. As such, any estimates calculated by this means should be treated as indicative.

4.4 Inhalation of aerosols

In addition to cloudshine the immersion of an individual in the plume of radioactive aerosols results in an exposure due to the inhalation of radioactive particulate.

Once inhaled this particulate may be absorbed or deposited in the body for a long period of time prior to being excreted, leading to a prolonged exposure or dose even once the plume has passed the individual.

A significant amount of research has gone into biokinetic models to simulate what happens to the radioactive particulate once it is inhaled into the human body, however, this is dependent on the physical and chemical form of the particulate, along with the age and gender of the receptor. This is discussed further in chapter 5.

The calculation of the inhalation dose to an individual receptor (E_{inh}) is one of the simpler exposure pathways. It is a function of c_i, the airborne radioactive concentration of each nuclide (in Bq m^{-3}), Br, the breathing rate of the receptor (in m^3 s^{-1}), t the exposure time to the plume (in s) and r_i, the radionuclide specific dose coefficient for a particular age group and aerosol particulate size (in Sv Bq^{-1}). The latter of which incorporate the biokinetics of the human body. This is summarised in the expression below:

$$E_{\mathrm{inh}} = tBr \sum_{i=0}^{n} c_i r_i \tag{4.5}$$

The inhalation pathway is only applicable to the radioactive particulate in the plume that can be deposited in the human body; in the case of gases which are exhaled shortly after inhalation the dose from the inhalation is usually discounted.

4.4.1 Re-suspension of deposited activity

In addition to the inhalation of radioactive particulates in the plume, there is the potential to disturb and re-suspend particulates which have been deposited on the ground and soil. This may be a result of many factors such as the wind, vehicle traffic, digging and farming activities. The re-suspended activity can then be inhaled by a receptor, as described earlier in section 4.4.

The level of re-suspension is dependent on the nature of the disturbance, depth of the deposited particulate in soil and particulate size, with smaller sized particulates being more easily mobilised.

Due to the nature of man-made disturbances, these can be difficult to model as they are situation and site specific, as such generic models only tend to exist for wind based disturbances.

The re-suspension is typically modelled using a re-suspension factor (k) which is a ratio of the radioactive particulate concentration in air due to re-suspension (in Bq m^{-3}), to surface radioactivity concentration (in Bq m^{-2}) [3]. However, due to the simplification of this method, selection or calculation of an appropriate re-suspension factor is key, to ensure the re-suspension pathway is suitably accounted for.

To date, wind re-suspension factors in the range of 10^{-2} to 10^{-10} m^{-1} have been measured.

4.5 Ingestion of contaminated food products

The deposition of radioactive particulates, as discussed in section 4.3, can lead to radioactive particulates entering the terrestrial food chain [7]. This can either be because of plant uptake from soil or direct deposition onto plants and crops. These contaminated plants and crops can then either be ingested by animals, with a portion of the radioactive contamination being transferred to food products such as milk, eggs or meat, or the plants or crops could be consumed directly by a human [8].

The concentration of radionuclides deposited onto plants or entering the food chain can be obtained by direct measurement or calculated depending on whether a prospective or retrospective assessment of the radioactive discharges is being undertaken.

Calculating deposition onto plants or crops can be undertaken using the methods discussed in section 4.3, however, a retention factor may need to be included to take account of the fraction of the deposition that adheres to the plants or crops.

The modelling of plant uptake and the transfer of radioactive contaminants through the food chain is more complex due to the number of physical and biological processes involved, and dependence on the physical and chemical properties of the radioactive contaminants. A summary of the key high level processes is provided in figure 4.3.

The selection of an appropriate model is key and depends on the application. Several models are available for use and have been developed for calculating the transfer of radionuclides through the chain as a result of both routine and accidental airborne discharges of radioactivity. These models tend to be compartmental in basis, with differing degrees of complexity, using a series of factors to determine the transfer of radioactivity between the various components of the food chain prior to

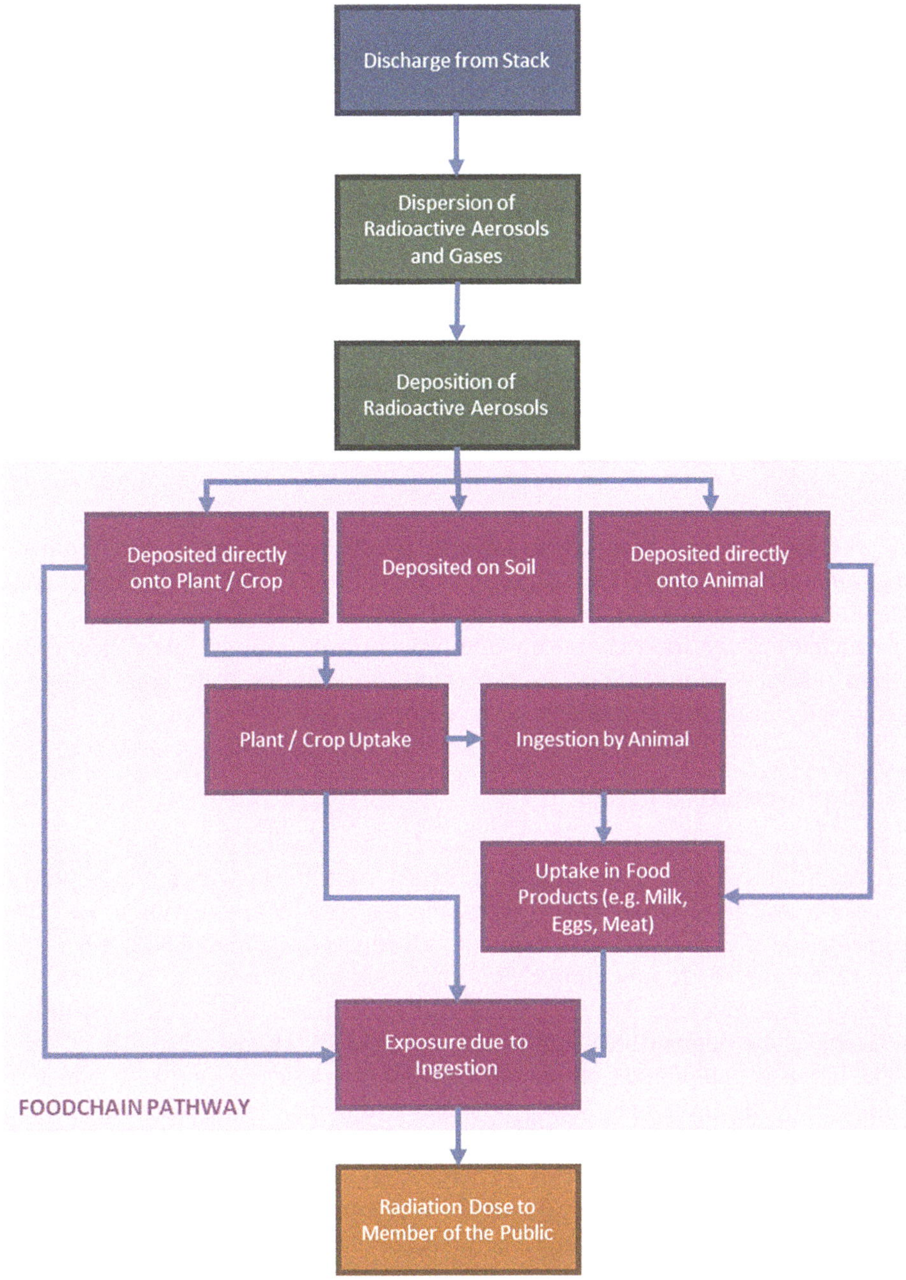

Figure 4.3. Key processes involved in the transfer of radioactive contaminants in the terrestrial food chain because of airborne radioactive discharges.

reaching the receptor. Compartmental models are generally appropriate for calculating the total amount of radioactivity transferred through the food chain, but are of limited use when considering time dependence.

Using these models, a value f_i (Bq kg^{-1}) can be obtained for the radionuclide concentration within a food product or plant/crop which is ingested by a human. The ingestion dose E_{ing} to a individual receptor is then calculated by multiplying k, the quantity of that particular food product or plant/crop ingested in a year (in kg), by f_i and I_i, a dose conversion coefficient (in Sv Bq^{-1}) and summing for all radionuclides, as seen below:

$$E_{ing} = k\sum_{i=0}^{n} f_i I_i \tag{4.6}$$

Care has to be taken selecting values of k, noting that people do not necessarily consume local agricultural products, and even common food products such as milk and vegetables tend to be imported and exported over large distances. It should be noted that although tracking the movement of food products is important the most exposed members of the public are usually those in close proximity to the source of the radioactive discharges.

It should be noted that the modelling of the movement of the radioactive contaminants through the food chain and subsequent calculation of the ingestion dose does not normally take account of the seasonal variation in uptake of radionuclides in the food chain, nor the reduction in radionuclide concentrations in food products as a result of food preparation. For instance, both boiling and frying food can reduce radionuclide concentration.

4.6 Representative person

When assessing the impacts to the members of the public, it is not practicable to evaluate the dose to each individual member. As such, the approach of a 'representative person' was introduced. This is defined as an 'individual receiving a dose that is representative of the more highly exposed individuals in the population' [9].

To determine the representative person, consideration needs to be given to the discharged radionuclides, the environment those radionuclides disperse within and the habits of the populations local to the source of the release.

The habits of the population include consumption rates of foodstuffs which may be affected by the discharges, occupancy rates within their living dwellings, and time they may spend in areas where they are subject to groundshine, cloudshine or inhalation of particulates.

Habit studies are regularly produced, and are the primary source of providing the required data, however, care should be taken as they can often be based on limited sample sizes, and as such are not always representative of a population group.

The most affected age group should be selected as the representative person.

Where habit data is not available, generic data is normally used. These are conservative in nature to ensure the assessed doses will encompass the more highly exposed individuals of the population. An example of generic habit data can be found in National Radiological Protection Board Publication W41 'Generalised Habit Data for Radiological Assessments' [10].

4.7 Novel pathways

In sections 4.1–4.5 we have discussed the more common exposure pathways. However, in certain cases 'novel' pathways may be identified as part of the determination of the representative person. This may include pathways such as:

- ingestion of novel foods;
- inadvertent ingestion of water, sediment, soil or sludge.

Where these are identified they should be considered as part of the assessment of the impacts associated with gaseous radioactive discharges.

4.8 Impacts to flora and fauna as a result of radioactive discharges

Over the last 20 years there has been increasing emphasis placed on the consideration of the impacts to the non-human environment as a result of radioactive discharges, rather than solely focusing on the human environment.

Although not the focus of this book, it should be noted that an increasing number of models have been developed with the aim of assessing the impacts and risks to various wildlife as a result of radioactive discharges. Much like the food chain model these are compartmental based.

The most common approach used to assess the impact to the non-human environment is called the 'Environmental Risk from Ionising Contaminants: Assessment and Management' or ERICA for short.

4.9 Summary

- 'Exposure pathways' are the routes that the radioactive aerosols and gas take before interacting with the human environment.
- The source, pathway, receptor model looks at where a discharge originates from, and the various pathways the contaminant travels through the environment prior to reaching a receptor. It is a useful tool for assessing the risks associated with radioactive discharges.
- Although the exposure pathways for infants, children and adults of both genders are broadly similar, the radiation dose as a result of the pathway may be different.
- There are four principal exposure pathways associated with airborne radioactive discharges and the public:
    - **Cloudshine**. Exposure an individual receives due to radiation from the immersion in the 'cloud' or plume of radioactive aerosols and gases.
    - **Groundshine**. Exposure an individual receives due to the build-up of radioactive particulates deposited on the surface of the ground and soil.
    - **Inhalation**. Exposure an individual receives due to inhalation of radioactive aerosols in the plume or re-suspended radioactive particulates deposited on the ground.
    - **Ingestion**. Exposure an individual receives due to ingestion of plants or crops directly contaminated with radioactive particulates, or food

products contaminated with radioactivity as a result of radionuclide transfer through the food chain.

In certain cases novel pathways may also exist.

- The radiation dose from the various exposure pathways can either be calculated from direct measurements of the radioactivity in the environment or modelled. This is dependent on whether a prospective or retrospective assessment of the radioactive discharges is being undertaken.
- Cloudshine is an applicable exposure pathway for both radioactive particulates and gases, whereas inhalation, ingestion and cloudshine are only relevant to radioactive particulates.
- Care must be taken when picking the appropriate models and factors for calculating the radiation exposure, as there are various limitations.
- When assessing doses to a member of the public the 'representative person' is used. This is defined as an 'individual receiving a dose that is representative of the more highly exposed individuals in the population'.
- In recent years there has been an increasing emphasis on considering the impacts of radioactive discharges both to the human environment and non-human environment. The most common approach for assessing the impacts to the non-human environment is called ERICA.

References

[1] Martin A, Harbison S, Beach K and Cole P 2012 *An Introduction to Radiation Protection* 6th edn (Boca Raton, FL: CRC Press)

[2] United Nations Scientific Committee on the Effects of Atomic Radiation 2000 *Sources and Effects of Ionizing Radiation: Sources* (New York: United Nations Publications)

[3] Smith J G and Simmonds J R 2009 *The Methodology for Assessing the Radiological Consequences of Routine Releases of Radionuclides to the Environment Used in PC-CREAM 08* HPA-RPD-058

[4] ICRP 1996 Conversion coefficients for use in radiological protection against external radiation ICRP publication 74 *Ann. ICRP* **26** 3–4

[5] Kocher D 1981 Dose-rate conversion factors for external exposure to photons and electrons *Oak Ridge National Laboratory* ORNL/NUREG-79

[6] Scott E 2003 *Modelling Radioactivity in the Environment* (Amsterdam: Elsevier)

[7] National Research Council 1995 *Radiation Dose Reconstruction and Epidemiologic Uses* (Washington, DC: The National Academies Press)

[8] IAEA 2001 Generic models for use in assessing the impact of discharges of radioactive substances to the environment *IAEA Safety Reports Series* No. 19

[9] ICRP 2006 Assessing dose of the representative person for the purpose of the radiation protection of the public ICRP publication 101a *Ann. ICRP* **36** 3

[10] Smith K R 2003 Generalised habit data for radiological assessments *National Radiological Protection Board* NRPB-W41

IOP Publishing

Airborne Radioactive Discharges and Human Health Effects
An introduction
Peter A Bryant

Chapter 5

Principles of dose and biological effects of radiation

In previous chapters we have covered the sources of radioactive discharges into the atmosphere and how they move in the environment, before ultimately interacting with the human population. But what impact do these discharges actually have on human health?

To understand this, we need to introduce the term 'dose' which is a measure of radiation exposure, and health impact. It is important not to confuse the terms 'dose' and 'radioactivity', as the latter refers to how unstable an isotope is, and the former is a measurement of the exposure of a body to ionising radiation, and although linked, one does not directly correlate to the other.

Radiation exposure can come from a number of sources and pathways, as discussed in chapters 2 and 4, and can result in an external dose, where ionising radiation originates from outside the human body, or internal dose where the ionising radiation originates from inside the human body, for instance due to the inhalation or ingestion of radioactive particulates [1].

In this chapter we will explore the principle of dose in further detail and the biological effects on the human body.

5.1 Absorbed dose

Radiation dose is defined in three different ways, the first of which is absorbed dose.

This is a measure of energy deposited in any medium, by any type of ionising radiation, per unit mass, and is summarised in figure 5.1 and equation (5.1):

$$D_{T,R} = \frac{Q}{m} \tag{5.1}$$

Q is the amount of energy deposited in the medium of mass m. The unit of absorbed dose is gray (Gy) with SI units of J kg^{-1}.

doi:10.1088/2053-2563/aafa6dch5

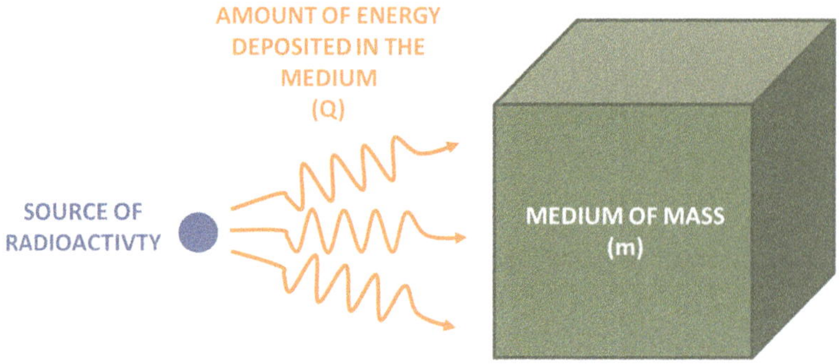

Figure 5.1. Diagrammatic representation of absorbed dose.

It can be seen from equation (5.1) that for a given Q and m the absorbed dose is the same regardless of what type of medium, such as tissue, absorbs the radiation or what type of radiation it is.

5.2 Equivalent dose

Although absorbed dose is useful, in order to quantify the biological effect of the radiation on a member of the public, two additional dose quantities are used.

Equivalent dose takes into account the fact that different types of radiation result in differing degrees of damage within a biological system [2]. For example, 0.05 Gy of alpha radiation can do as much biological damage as 1 Gy of gamma radiation. Or to put it another way, alpha radiation is 20 times more ionising or damaging than gamma radiation.

In order to take account of this 'radiobiological effectiveness' the absorbed dose for a particular type of radiation is multiplied by a radiation weighting factor (w_R). This is summarised in the equation below:

$$H_T = \sum_R W_R D_{T,R} \tag{5.2}$$

where H_T is the equivalent dose to a tissue. Equivalent dose is measured in units of the Sievert (Sv), and has the same SI units as absorbed dose of J kg^{-1}.

A summary of the different radiation weighting factors is presented in table 5.1:

5.3 Effective dose

The last dose quantity is called the effective dose and takes account of the fact that different organs and tissues within the human body have differing sensitivities to ionising radiation. This is particularly important in situations where the human body is not uniformly exposed.

The effective dose to a tissue is obtained by taking the equivalent dose and multiplying it by a tissue weighting factor w_T which relates to the organs/tissues under consideration (table 5.2) [3]:

Table 5.1. Summary of radiation weighting factors, (w_R).

Radiation type	Radiation weighting factor
Alpha	20
Neutron	A continuous w_R function is chosen with a maximum of 20 at ~1 MeV
Beta, x-ray and gamma	1

Table 5.2. Summary of tissue weighting factors, (w_R).

	Tissue weighting factor	
Tissue	w_T	Σw_T
Bone-marrow (red), colon, lung, stomach, breast, remainder tissues[a]	0.12	0.72
Gonads	0.08	0.08
Bladder, oesophagus, liver, thyroid	0.04	0.16
Bone surface, brain, salivary glands, skin	0.01	0.04
Total		1.00

[a] Remainder tissues: adrenals, extra thoracic region, gall bladder, heart, kidneys, lymphatic nodes, muscle, oral mucosa, pancreas, prostate, small intestine, spleen, thymus, uterus/cervix.

The total effective dose to an individual (E) is used to express the overall health detriment as a summation of several doses of different radiation types to different tissues:

$$E = \sum_T W_T H_T \tag{5.3}$$

Effective dose is also measured in units of the Sievert.

In the following chapters the term dose will generally be taken to mean either equivalent dose or effective dose depending on the context.

5.4 Basic human physiology

As previously mentioned, exposure to ionising radiation may result in an external dose, or internal dose. In the case of the latter the ionising radiation originates from inside the human body, for instance due to the inhalation or ingestion of radioactive particulates.

In the case of external dose, the exposure or dose is usually short lived, as once the individual moves away from the source of the radiation the exposure stops. However, in the case of an internal dose the exposure time is more complicated, and may last for prolonged periods depending on the physical and chemical form of the radionuclide.

5.4.1 Inhalation and the respiratory system

As discussed in chapter 4, one of the dominant exposure pathways to a member of the public from radioactive discharges is due to the inhalation of radioactive particulates.

The diameter of the radioactive particulate dictates how deep into the airways the particulate penetrates, for instance:

- aerosols with a medium diameter of 10 μm can penetrate the respiratory system beyond the larynx;
- aerosols with a medium diameter of 4 μm can penetrate to the unciliated airways of the lungs (alveolar region);
- in addition, a medium diameter of 2.5 μm corresponds to the respirable fraction for children and infants.

Once the particulate penetrates the respiratory system the material is either deposited or exhaled. Once deposited, the behaviour of the particulate is dependent on its solubility and chemical form. For instance, the more soluble the materials are the more readily they are absorbed into the bloodstream, whereas insoluble material may persist in the lungs for many months.

In addition, the inhalation of radioactive gases can result in an internal dose due to the gases passing freely into the lungs and entering the bloodstream. This is also dependent on the solubility of the gases, however, as they spend only a short period of time in the airways before they are exhaled, the levels of radioactive material entering the bloodstream are usually lower than from particulates.

Inhalation of radioactive particulate or gases can therefore result in radioactive material remaining in the lungs for long periods of time, where the material can irradiate the lung until it decays or is exhaled, or is transported by the bloodstream to other parts of the body.

5.4.2 Ingestion and the digestive system

One of the other dominant exposure pathways discussed in chapter 4 was the ingestion of contaminated food products. Once ingested, the food and contaminants are converted into a form suitable for the production of heat and energy and the materials necessary for the growth and repair of tissues.

The larger molecules in the food and soluble contaminants are broken down by digestive enzymes in the digestive tract and absorbed into the bloodstream via the liver.

The unabsorbed food and insoluble contaminants are passed out of the body as either solid waste (faeces) or urine. During their passage the insoluble contaminants will irradiate the digestive tract and the large intestine.

5.4.3 Radionuclides in the circulatory system

Once absorbed into the bloodstream, whether from the inhalation or ingestion of radioactive contamination, the soluble contaminants can be transported by the circulatory system to other parts of the body [1]. Depending on the chemical form of the contaminant, the radionuclide may become concentrated in specific organs or tissues, where it will irradiate the organ/tissue until it decays or is excreted.

For example, the thyroid gland readily absorbs iodine, found in many foods, and converts it into thyroid hormones. As such when a radioactive isotope of iodine such as iodine-131 is inhaled or ingested it can readily concentrate in the thyroid, and stay there for a prolonged period of time. It is for that reason that in the event of a nuclear reactor accident the population in the vicinity of the reactor may be given tablets containing stable iodine, such that the thyroid gland absorbs the stable iodine and does not need to absorb any more iodine from the inhaled or ingested contaminants.

5.4.4 Biological half-life

The rate of excretion from the body is radionuclide specific and is measured using an effective biological half-life.

The rate of decrease of radiation exposure from a particular radionuclide (or effective half-life) is therefore a function of radioactive half-life (as discussed in chapter 1) and biological half-life, as expressed below:

$$\frac{1}{t_e} = \frac{1}{t_r} + \frac{1}{t_b} \tag{5.4}$$

where t_e is the effective half-life, t_r is the radioactive half-life and t_b is the biological half-life.

5.5 Cell biology

All organs and tissues within the human body are made up of cells, a diagram of which is presented in figure 5.2. These cells readily reproduce to compensate for the cells that die. This primarily takes place via a process called mitosis, where the cell divides into two new cells each identical to the original.

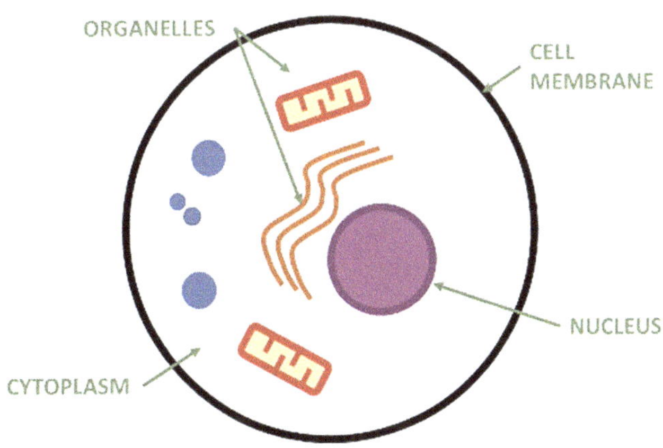

Figure 5.2. Diagrammatic representation of a 'typical' human cell.

Each cell contains a nucleus, which consists of chromosomes made up of genes. These genes include deoxyribonucleic acid (DNA), which comprises the information used to determine the characteristics of any progeny cells produced by mitosis.

Whether due to an external dose or internal dose the primary means by which the ionising radiation interacts with a human cell is the same, with the only difference being the location of the cell impacted and exposure period (as discussed previously).

When the radiation interacts with the atoms within the cell, it causes ionisation (as discussed in chapter 1). This ionisation can result in chemical changes within the cell, which can either damage the chromosomes or DNA, or disrupt the delicate chemical balance allowing the cell to perform its intended function. Depending on the severity of the change and the number of cells impacted these changes can in turn manifest themselves as deterministic effects (such as radiation sickness), or in the longer term stochastic effects (such as cancer).

This is discussed in further detail in the subsequent sections.

5.6 Deterministic effects

Deterministic effects results from acute high doses of ionising radiation to either a part of the body or the whole body, leading to cell damage or death [3]. It is characterised by a dose threshold, at which an effect is observed and above which the severity of the effect or damage increases with increasing dose, as summarised in figure 5.3.

Examples of these effects include skin reddening (radiation burns), hair loss, cataracts and radiation sickness (nausea, vomiting and diarrhoea) and the thresholds have been identified from the substantial data from:

- Japanese A-bomb survivors, and
- medically exposed groups.

Examples of typical dose thresholds are provided in table 5.3 [4].

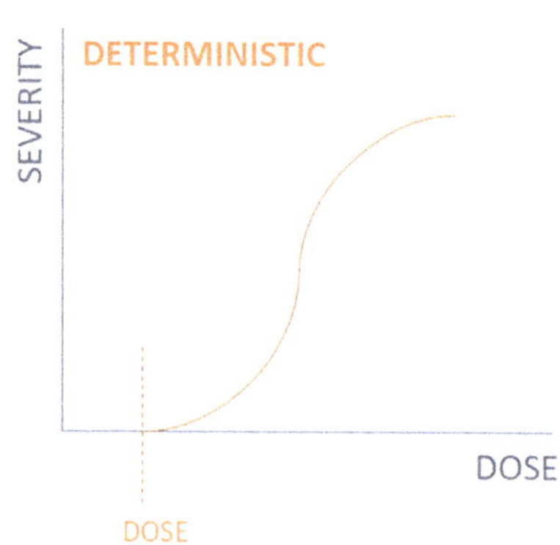

Figure 5.3. Characteristics of a deterministic effect.

Table 5.3. Dose thresholds for non-stochastic effects.

Dose (Sv)	Typical early effect
50	Severe gastrointestinal and central nervous damage; death rapidly ensues
10	Observable damage to exposed organ; death probably within weeks, if whole body exposed
3	50% chance of death if whole body exposed
1	Some signs of radiation sickness if whole body exposed
0.5	Detectable changes in numbers of blood cells
0.2	Detectable chromosome changes in blood cells

5.7 Stochastic effects

Stochastic, or as they are also referred to probabilistic effects, take account that even the smallest quantity of ionising radiation exposure can be said to have a finite probability of causing an effect, such as cancer in the individual or genetic damage that may be passed down to future generations. Unlike deterministic effects there is no threshold associated with stochastic effects, and it is the probability of an effect occurring that increases with dose, rather than the severity of the effect. The established model of probabilistic effects is a linear relationship between effective dose and increased risk of an effect as seen in figure 5.4 below.

The dotted line denotes that there is limited available data at low chronic doses, and as such the relationship is conservatively assumed to be linear. Future data from the following sources will be used in the future to continue to develop and refine the model further.

- Studies of occupationally exposed groups (such as nuclear power plant workers or medical professionals).
- Studies of the former Soviet Union (Chernobyl) and Japan (Fukushima).

The probabilistic effects of ionising radiation are due to chemical damage of DNA caused by the ionisation, as discussed in section 5.5. In some cases an ionising particle or ray will cause a double strand break in the DNA which if incorrectly repaired can lead to cancer or genetic effects.

Current scientific consensus assumes a low dose fatal cancer risk of 4% per Sv.

5.8 Dose conversion coefficients

As seen in chapter 4 when modelling the dose to a member of the public from a particular exposure pathway dose conversion coefficients are used. These are used to convert the uptake of radioactivity by a person into an effective dose, as seen below:

$$E_i = A_i e_i$$

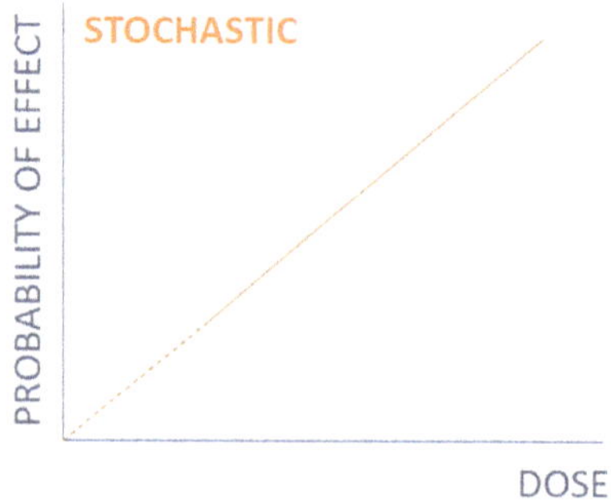

Figure 5.4. Characteristics of a stochastic effect.

where E_i is the effective dose from radionuclide i (in Sv), A_i is the activity of radionuclide i taken into the individual's body (in Bq) and e_i is the dose conversion coefficient (in Sv Bq^{-1}).

The dose conversion coefficient is a measure of the radiotoxicity of the radionuclide for a given uptake pathway. It incorporates the radiation and tissue type weighting factors as well as biokinetic modelling of the human body.

The dose conversion coefficients depend on several factors:

- The uptake pathway, e.g. inhalation, ingestion or wounding. Different dose conversion coefficients are given for each uptake pathway.
- The physical characteristics of the dose recipient. Dose conversion coefficients are given for reference persons (adult, children and infants) with average physical characteristics.
- The chemical state of the contaminants. Inhalation dose conversion coefficients are given in three categories, 'F', 'M' and 'S' which denotes the solubility and different absorption rates of the material deposited within the lung into the bloodstream.
- If the contaminants are in the form of an aerosol, the size of the aerosol particles determine how far into the lung they can penetrate and hence how much activity is absorbed by the blood. The aerosol particle size is characterised by the activity median aerodynamic diameter (AMAD): 50% of the activity in the aerosol is carried by particles which are larger than the AMAD. Inhalation dose coefficients are given for two values of AMAD: 1 μm and 5 μm. For aerosols with 1 μm AMAD, 50% of the activity penetrates to the deep lung. At 5 μm AMAD, most activity is deposited in the upper respiratory tract.

5.9 Doses in the context of radioactive discharges

In chapter 2 we noted that the average worldwide background radiation dose was 2.4 mSv yr^{-1} or 0.002 Sv yr^{-1}, or which only 0.01% or 0.000 0002 Sv yr^{-1} came from routine radioactive discharges.

For populations neighbouring a nuclear site such as a nuclear power station the annual doses from routine discharges are slightly higher, typically around

~0.000 02 Sv yr^{-1}, which is still very low in comparison to the background levels of radiation from natural sources such as radon.

In both cases the doses associated with the discharges are well below the deterministic thresholds presented in section 5.6.

The typical increase in fatal cancer risk to a member of the public living near a nuclear site from the associated radioactive discharges is ~0.000 08% per year, this is far lower than the risk of cancer from smoking or routinely drinking alcohol.

However, what about non-routine discharges such as in the case of a nuclear accident like Fukushima?

Of the ~200 000 residents living in the vicinity of Fukushima Daiichi no deterministic health effects were found by the end of May 2011. Government health checks of residents evacuated from three municipalities showed that approximately two-thirds received an external radiation dose below 1 mSv yr^{-1}, 98% were below 5 mSv yr^{-1}, and ten people were exposed to more than 10 mSv [6].

Although the potential fatal cancer risk to the neighbouring population from the accident is higher than from routine discharges, the level of risk is still broadly equivalent to the risk associated with background levels of radiation in the areas of 3 mSv yr^{-1}, with the largest increase in fatal cancer risk of ~ 0.132%.

5.10 Summary

- 'Dose' is a measure of radiation exposure, and health impact.
- Radiation exposure can result in an external dose, where the ionising radiation originates from outside the human body, or internal dose where the ionising radiation originates from inside the human body, for instance due to the inhalation or ingestion of radioactive particulate.
- Radiation dose is defined in three different ways:
    - **Absorbed dose** Measure of energy deposited in any medium, by any type of ionising radiation, per unit mass. It has the units Gy or J kg^{-1}.
    - **Equivalent dose** Takes into account the fact that different types of radiation result a differing degrees of damage within a biological system. It has units of the Sv, and has the same SI units as absorbed dose of J kg^{-1}.
    - **Effective dose** Takes account that different organs and tissues within the human body have differing sensitivities to ionising radiation. The same units as equivalent dose.
- The inhalation or ingestion of radioactive materials may result in prolonged periods of exposure depending the physical and chemical form of the radionuclide.
- In the case of the inhalation of radioactive aerosols the diameter of the radioactive particulate dictates how deep into the airways the particulate penetrates, for instance:
    - Aerosols with a medium diameter of 10 μm can penetrate the respiratory system beyond the larynx.

- Aerosols with a medium diameter of 4 μm can penetrate to the unciliated airways of the lungs (alveolar region).

Once the particulate penetrates the respiratory system the material is either exhaled or deposited, where depending on its solubility it may be absorbed into the bloodstream.

- The inhalation of radioactive gases can also result in an internal dose due to the gases passing freely into the lungs and entering the bloodstream.
- Ingestion of contaminated food products may result in the contaminant either being passed out of the body as waste (faeces or urine) or passed into the bloodstream. This is dependent on its solubility.
- Once absorbed into the bloodstream, radioactive contamination can be transported by the circulatory system to other parts of the body where it may become concentrated in specific organs or tissues, where it will irradiate the organ/tissue until it decays or is excreted.
- The biological half-life is a term used to measure the rate of excretion of a radionuclide from the human body. The rate of decrease of radiation exposure from a particular radionuclides is a function of both the radioactive half-life and biological half-life.
- When radiation interacts with a human cell it causes ionisation which can result in chemical changes within the cell, which can either damage DNA, or disrupt the delicate chemical balance allowing the cell to perform its intended function.
- Acute, high dose exposures can result in deterministic effects, these are characterised by a dose threshold, at which an effect is observed and above which the severity of the effect or damage increases with increasing dose.
- Chronic, low dose exposures can result in stochastic or probabilistic effects. Unlike deterministic effects there is no threshold associated with stochastic effects, and it is the probability of an effect occurring that increases with dose, rather than the severity of the effect.
- Dose conversion coefficients are used to convert the uptake of radioactivity by a person into an effective dose taking account of the radiotoxicity of the radionuclide for a given uptake pathway, as well as biokinetic modelling of the human body.
- The doses associated with routine discharges of radioactivity into the environment are very low in comparison to the background levels of radiation from natural sources such as radon. The doses associated with the discharges are well below the deterministic thresholds and far lower than the risk of cancer from smoking or routinely drinking alcohol.

References

[1] Environment Agency 2012 *Principles for the Assessment of Prospective Public Doses arising from Authorised Discharges of Radioactive Waste to the Environment.*

[2] Knoll G F 2000 *Radiation Detection and Measurement* 3rd edn (New York: Wiley)

[3] ICRP 2007 The 2007 recommendations of the international commission on radiological protection ICRP publication 103 *Ann. ICRP* **37**
[4] Martin A, Harbison S, Beach K and Cole P 2012 *An Introduction to Radiation Protection* 6th edn (Boca Raton, FL: CRC Press)
[5] Health & Safety Executive 1988 *The Tolerability of Risk from Nuclear Power Stations* (London: HMSO)
[6] World Nuclear Association *Fukushima: Radiation Exposure* http://world-nuclear.org/information-library/safety-and-security/safety-of-plants/appendices/fukushima-radiation-exposure.aspx [accessed 23 March 2018]

Chapter 6

Environmental monitoring systems

In the previous chapters we explored the sources of radioactive discharges, how they move in the environment, and how we assess the potential impacts to members of the public and flora and fauna. But how do we confirm these predictions are correct?

Depending on the level of discharges it is common practice to sample and monitor radioactive discharges, both during planned discharges and under accident conditions. This can be undertaken in one of two ways. Sampling at the source of the discharges (in process sampling) or sampling of the environment.

In this chapter we will explore some of the most commons techniques for the sampling and measuring of radioactive discharges.

6.1 Sampling radioactive discharges at source

The most efficient way to sample what discharges have been released into the environment is to do so at the source. This can either be done by sampling at the point of discharge such as the stack, or earlier in process within the facility and extrapolating the end discharges.

The measured discharges can then be used to assess the discharges against any proposed limits imposed by the relevant national regulator, along with reporting the discharge levels to the national regulator [1].

Depending on whether you are measuring discharges of aerosols or gases, the techniques used usually consist of taking a sample of the discharge and subsequently measuring the radioactivity in the sample.

This is summarised in figure 6.1, where a pump draws a portion of the discharge from the stack via a stack monitor, where a sample is collected. Depending on the type of stack monitor, radionuclides and physical form of the discharge (e.g. aerosol or gas), a real-time measurement of the discharge can be made, or a sample retained to subsequently be sent off to a lab for analysis.

doi:10.1088/2053-2563/aafa6dch6

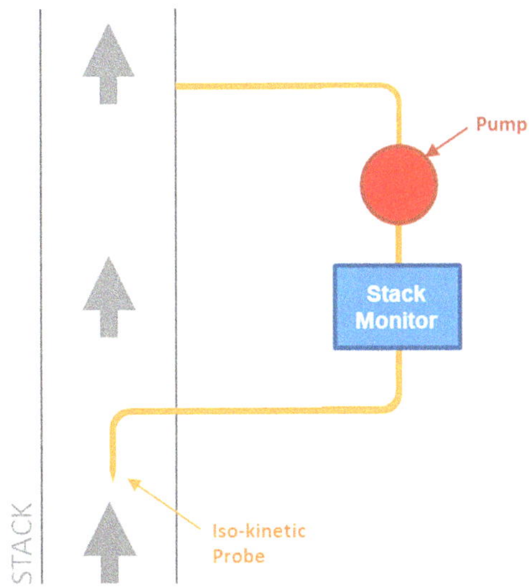

Figure 6.1. Diagrammatic representation of stack sampling arrangement.

Discharges are usually recorded/measured in five different categories:

- aerosols (excluding iodine);
- iodine;
- noble gases;
- tritium;
- carbon-14.

In the next few sections we will explore the key methods for sampling and measuring these types of radioactive discharges to the atmosphere at the point of discharge.

6.1.1 Aerosol sampling

When sampling aerosols there are two important factors to consider. First, the design of the discharge pipework and location of the sampling equipment, and second, obtaining a representative sample.

The pipework used to carry the discharge should minimise any sharp bends, as these affect the flow of the discharge. Aerosols of a larger diameter may deviate from the streamlines as they go around a bend, leading to the aerosols potentially impacting on the surfaces of the pipework and a build-up of contamination [2].

To help predict whether an aerosol will follow the flow of a gas as it turns a bend the Stokes number (St) is used. This value is a function of velocity of the discharge (u_0), diameter of the pipework (l_0) and relaxation time (t_0), where the relaxation time is small for small aerosols, namely those that follow the streamlines more easily, and large for larger aerosols which are more likely to deviate from the streamlines.

$$St = \frac{t_0 u_0}{l_0} \tag{6.1}$$

It is for this reason it is also important that the sampling point does not occur near any bends. This is to ensure that the discharge is appropriately mixed prior to being measured.

The sampling point should also be designed to ensure representative sampling of the aerosols. To do this isokinetic sampling is used. As seen in figure 6.2, when the velocity of the discharge is less than the sampling flow ($u_1 < u_2$), large aerosol particles are excluded from the sample and smaller aerosol particles are preferentially captured.

Meanwhile, if the velocity of the discharge is greater than the sampling velocity ($u_1 > u_2$), then small particles will follow the streamlines around the sampling point leading to preferential collection of larger aerosol particles.

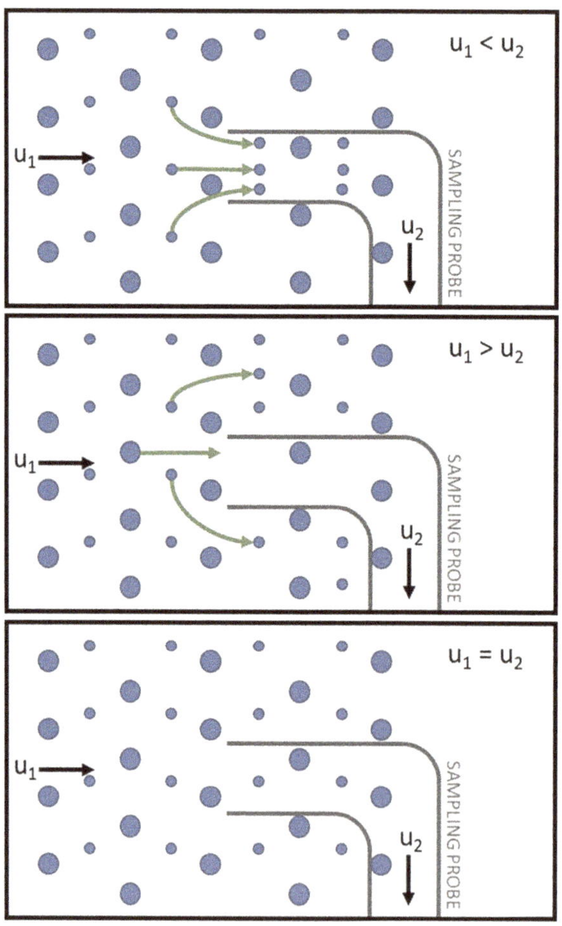

Figure 6.2. Diagrammatic representation of isokinetic sampling.

For this reason, it is important to match the velocity of the discharge with the sampling velocity ($u_1 = u_2$) thus ensuring that a representative sample is obtained.

As previously mentioned, once sampled, the radioactivity in the sample is either measured by the stack monitor, or a sample retained to subsequently be sent off to a lab for analysis. Stack monitors are therefore broken down into two categories, active samplers or passive samplers.

In both cases the sample of the discharge is drawn through a semi-permeable paper barrier (filter paper) which collects any aerosols within the sample. For passive samplers the filter papers are periodically collected from the stack monitor and sent off to a laboratory for analysis. However, in the case of an active sampler a radiation detector is built into the stack monitor so a real-time measurement of the radioactivity in the discharge can be made. In the case of an active sampler the radiation detector assembly will consist of a shielded unit housing the filter and detector. This is to remove any interference from other sources of radioactivity [2].

Using the measured radioactivity in the sample (S_A) in Bq, the flow rate of the discharge in the stack (F_S) in m^3 s^{-1} and flow rate of the sample through the stack monitor (F_M) in m^3 s^{-1}, the discharged activity (D_A) in Bq can be calculated using the equation below:

$$D_A = \left(\frac{F_S}{F_M}\right) S_A \tag{6.2}$$

Depending on the type of facility, you can find either active or passive samplers or even a combination of the two. In the latter case the passive sampler is likely to be used for re-assurance monitoring in the event the active sampler is not operational or the results from the active sampler require additional validation (figure 6.3).

6.1.2 Sampling of iodine

Radioactive halogens such as iodine-131 are specifically measured (where discharged) due to their importance from an environmental point of view. This is due to the

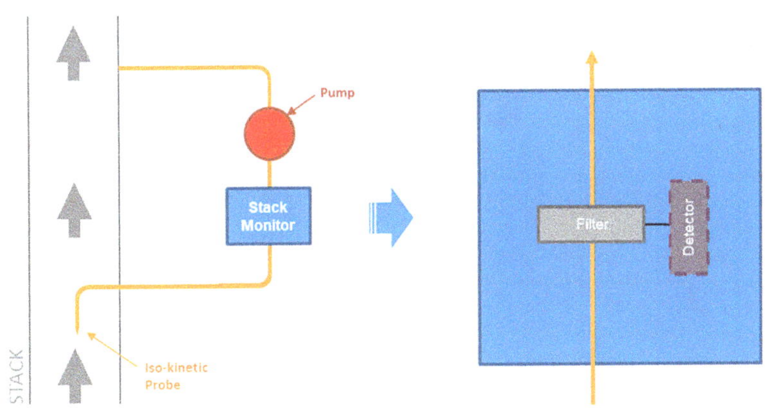

Figure 6.3. Diagrammatic representation of an active stack sampler.

radiotoxic properties, namely the potential to be inhaled or deposited on pasture grasses, followed by the ingestion by animals such as cows and the subsequent consumption of contaminated milk and dairy products by humans [4, 5]. Once in the human body, iodine-131 is taken up by the thyroid where it may remain for a long period of time leading to a prolonged internal dose.

Radioactive iodine discharges may be in the form of both aerosols or as a gas; for this reason the stack monitoring arrangements for iodine are broadly similar to those described for aerosols, however, in order to collect a sample of gaseous iodine an appropriate solid absorbent material such as a charcoal filter arrangement is needed. Both active and passive iodine monitoring of iodine is possible.

6.1.3 Sampling of noble gases

Due to the physical form of noble gases such as krypton and xenon isotopes and argon-41, it is not possible to measure the radioactivity using the same methods as aerosols or iodine where they are discharged.

A sample of the noble gases is taken and passed through a fixed volume calibrated measurement chamber, with an appropriate radiation detection system.

The noble gas sample is collected and measured after the removal or any particulate material and, which is usually collected for separate analysis. This is summarised in figure 6.4.

The measurement of the radioactivity in the noble gas sample is continuously monitored and used to calculate the noble gas radioactivity discharged from the facility using equation (6.2).

6.1.4 Sampling of tritium and carbon-14

Both tritium and carbon-14 have reasonably long half-lives, and are highly mobile in the environment due to their chemical properties. For this reason, specific considerations are required for the management of their discharges.

A common problem with tritium and cabon-14 is that they are both difficult to measure, since they are both weak beta emitters and do not emit gamma radiation.

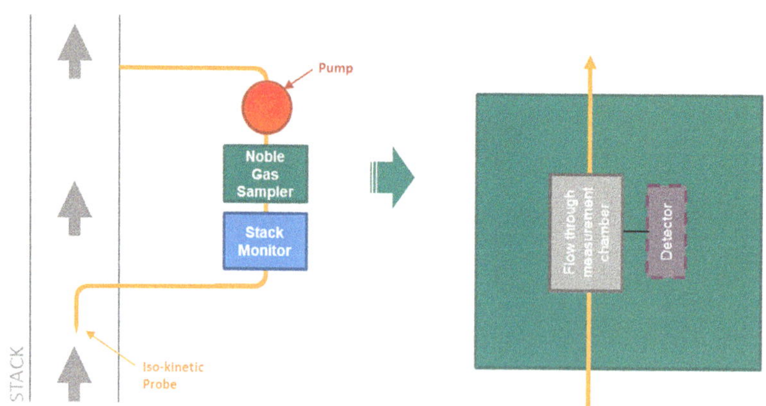

Figure 6.4. Diagrammatic representation of a noble gas sampler.

For this reason, it is not possible to measure tritium or carbon-14 discharges in real time [6].

Instead, a series of bubblers are commonly used to collect samples for analysis by an appropriate analytical technique such as liquid scintillation counting in a laboratory. An example of a bubbler system is shown in figure 6.5.

The sample collection system is made up of a series of bubbler traps to prevent any loss of sample through the first trap. The bubbler may also include a furnace and/or catalyst to enable both organic and inorganic forms of tritium or carbon-14 to be collected. For instance, in figure 6.5 the first set of vials would collect any tritium oxide (HTO) from the discharge, and the second set of vials would collect any tritium gas (HT) after it has been oxidised [7].

The vials would be periodically sent off for analysis to determine the organic and inorganic tritium or carbon-14 concentrations in the sample in Bq. This can be used to calculate the tritium or carbon-14 radioactivity discharged from the facility using equation (6.2).

6.2 Sampling of radioactive discharges in the environment

In addition to the sampling of the discharges at source, the national regulators often require facility operators (such as nuclear facilities) to assess the potential radiological impacts through regular monitoring of the environment surrounding the facility [8].

In addition, national radioactivity monitoring programmes may be carried out by state bodies.

Both environmental monitoring programmes may include measurement of radiation levels (called 'ambient dose') as well as radioactivity levels in air, soil, vegetation, food and water [9].

Measurement of ambient dose is usually undertaken either via the use of passive monitors, similar to the badges worn by radiation workers for measuring dose, located at fixed locations around the perimeter of the site. These are then periodically (for instance every three months) sent to a lab for development to determine the

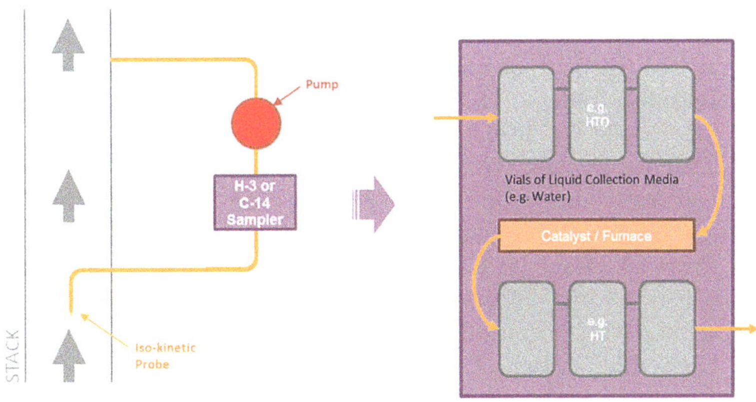

Figure 6.5. Diagrammatic representation of a tritium sampler.

level of radiation it has been exposed to in that period. Alternatively, direct measurements of the ambient dose can be made using a radiation detector, as described in chapter 7.

It should be noted that due to the physical properties of noble gases the dose to the human environment is predominantly external to the human body and as such the contribution to the environmental impacts is measured as part of ambient dose.

Measurement of radioactivity in air is usually done via environmental aerosol samplers, whereas, due to the difficulties in measuring tritium and carbon-14, as discussed earlier, the environmental impacts from tritium and carbon-14 are usually determined from measurements of radioactivity in soil, vegetation, food and water. This is described further in sections 6.2.1 and 6.2.2.

Doses to people living near the facilities are then assessed using results obtained from the radionuclide concentrations in the samples, ambient dose rates, aerosol measurements and information on the habits of people living near the facility [10].

From year to year, doses to people can vary, mostly because of changes in radionuclide concentrations and external dose rates. Data and information regarding changes in habits, in particular food consumption, can also contribute to the variation in the estimation of dose between years.

6.2.1 Environmental aerosol samplers

Environmental aerosol samplers can be broadly broken down into three different types:

- **Total aerosol samplers**. These samplers collect all the aerosol materials they come in contact with regardless of particle size, type and composition. Although simple in design and construction (e.g. may not require the use of a pump) such as a tacky shade, they are often unrealistic and problematic due to factors such as low concentrations, wide particle size distributions and various wind speeds.
- **Sampling for human health effects**. These types of sampler try to mimic the human respiratory system noting that the smaller the size of aerosol particle the deeper into the human body the aerosol may be deposited. This is summarised via the following international conventions below:
    - **Inhalable fraction**. This is the mass fraction of total airborne particulate inhaled via the nose and mouth.
    - **Thoracic fraction**. This is the mass fraction of airborne particulate that penetrates the respiratory system beyond the larynx. In general, the median diameter corresponds to 10 μm.
    - **Respirable fraction**. This is the mass fraction of airborne particulate that penetrates the unciliated airways of the lungs (alveolar region). In general, the median diameter corresponds to 4 μm.
    - **High risk fraction**. This is the mass fraction of airborne particulate particularly relevant to children and infants with a medium diameter that corresponds to 2.5 μm.

In the US a slightly different convention is used where there are only two conventions, PM10 and PM2.5, where the medium diameters correspond to 10.6 μm and 2.5 μm, respectively.

An example of a human health effect environmental aerosol sampler is the Graseby Anderson high flow rate sampler. Unlike total aerosol samplers, these types of samplers are more complex requiring the use of a pump.

- **Sampling for particle size distribution**. These are used to divide an aerosol sample into a range of particle sizes. The air is drawn into the top of the device, where the particles are exposed to a number of stages involving an aperture and an impaction plate. Each successive stage incorporates a higher particle velocity and a narrower gap between the aperture and the plate. This makes it progressively more difficult for large particles to navigate, resulting in each place collecting particles in specific size ranges, from large particles at the top to small particles at the bottom. The impact plates are usually sticky to prevent the particles rebounding off the plates. An example of particle size distribution aerosol sampler is the cascade impactor.

Depending on the country, a wide range of different environmental aerosols samplers are used. The selection of the sampler is used based on a risk informed point of view.

6.2.2 Sampling of environmental media

As previously mentioned, in addition to ambient dose and airborne aerosol concentrations, radioactivity levels are also measured in samples of soil, vegetation, food and water [11].

Samples are selected that may be affected by discharges to the air, although where food availability is limited, environmental indicator materials such as grass and soil are monitored.

The types of samples collected are chosen based on the location of the facility and habits of people living near the facility. This includes information on the main components of their diet such as milk, meat and cereals, and products most likely to be contaminated by the discharges, such as leafy green vegetables or soft fruit. Foods such as mushrooms and honey, which are known to accumulate radionuclides, may also be sampled.

In general, the samples are analysed for the presence of tritium, carbon-14, iodine isotopes and gamma emitting radionuclides, however, analyses for additional radionuclides may be carried out depending on the discharges of the local facilities.

6.3 Summary

- Sampling and monitoring of radioactive discharges and the environment can be undertaken via sampling at the source of the discharges (in process sampling) or sampling of the environment.

- Sampling and monitoring at the source of the discharge allows the measured discharges to be assessed against any limits imposed by the relevant national regulator, along with reporting the discharge levels to the national regulator.
- Discharges are usually recorded/measured in five different categories:
    - aerosols (excluding iodine);
    - iodine;
    - noble gases;
    - tritium;
    - carbon-14.
- Depending on whether you are measuring discharges of aerosols or gases, the techniques used usually consist of taking a sample of the discharge and subsequently measuring the radioactivity in the sample.
- When designing the sampling arrangement for aerosols it is important that the sampling point does not occur near any bends. This is to ensure that the discharge is appropriately mixed prior to being measured. In addition, the sampling point should be designed to ensure representative sampling of the aerosols. To do this isokinetic sampling is used.
- Stack monitors for aerosols may be active samplers or passive samplers. Active samplers have an inbuilt radiation detector allowing a real-time measurement of the activity in the discharge to be measured, whereas passive samplers require the sample to be periodically collected from the stack monitor and sent off to a laboratory for analysis.
- Iodine-131 is specifically measured (where discharged) due to its importance from an environmental point of view. Radioactive iodine discharges may be in the form of both aerosols or as a gas, for this reason the stack monitoring arrangements for iodine are broadly similar to those described for aerosols.
- Due to the physical form of noble gases such as krypton and xenon isotopes and argon-41, a different sampling arrangement is used compared to aerosols. This involves the sample of the noble gases being passed through a fixed volume calibrate measurement chamber, with an appropriate radiation detection system.
- Both tritium and carbon-14 are highly mobile in the environment and are both difficult to measure and as such special sampling arrangements are required. This involves the use of a series of bubblers to collect samples for analysis by an appropriate analytical technique such as liquid scintillation counting.
- In addition to the sampling of the discharges at source, the national regulators often require facility operators (such as nuclear facilities) to assess the potential radiological impacts through regular monitoring of the environment surrounding the facility. In addition, national radioactivity monitoring programmes may be carried out by state bodies.
- Sampling and monitoring of radioactive discharge in the environment allows a retrospective assessment of the doses to people living near the relevant facilities to be undertaken. This is based on results obtained from the radionuclide concentrations in the samples, ambient dose rates, aerosol measurements and information on the habits of people living near the facility.

- Measurement of ambient dose is usually undertaken either via the use of passive monitors, similar to those badges worn by radiation workers for measuring dose, located at fixed locations around the perimeter of the site.
- Measurement of radioactivity in air is usually done via environmental aerosol samplers. Environmental aerosol samplers broadly break down into three different types:
  - **Total aerosol samplers**. These samplers collect all the aerosol materials they come in contact with regardless of particle size, type and composition.
  - **Sampling for human health effects**. These types of samplers try to mimic the human respiratory system noting that the smaller the size of aerosol particle the deeper into the human body the aerosol may be deposited.
  - **Sampling for particle size distribution**. These are used to divide an aerosol sample into a range of particle sizes.
- Radioactivity levels are also measured in samples of soil, vegetation, food and water. Samples are selected that may be affected by discharges to the air, although where food availability is limited, environmental indicator materials such as grass and soil are monitored.

References

[1] Martin A, Harbison S, Beach K and Cole P 2012 *An Introduction to Radiation Protection* 6th edn (Boca Raton, FL: CRC Press)

[2] Luykx F and Fraser G 1983 *Radioactive Effluents From Nuclear Power Stations and Nuclear Fuel Reprocessing Plants in the European Community* (Luxembourg: Commission of the European Communities)

[3] Knoll G F 2000 *Radiation Detection and Measurement* 3rd edn (New York: Wiley)

[4] Hitachi 2017 *UK ABWR Generic Design Assessment Approach to Sampling and Monitoring* Ver 0

[5] ATSDR 2002 *Radiation Exposure from Iodine 131* (Atlanta, GA: US Department of Health and Human Services)

[6] IAEA 2004 Management of waste containing tritium and carbon-14 *Technical Reports Series No. 421* (Austria: IAEA)

[7] IAEA 1991 Safe handling of tritium review of data and experience *Technical Reports Series No. 324* (Austria: IAEA)

[8] ENSREG *Environmental Monitoring* http://ensreg.eu/nuclear-safety/environmental-monitoring [accessed 15 July 2018]

[9] IAEA 2005 Environmental and source monitoring for purposed of radiation protection *IAEA Safety Standards Series No. RS-G-1.8* (Austria: IAEA)

[10] Sellafield 2017 *Monitoring our Environment Discharges and Environmental Monitoring Annual Report 2016* Nuclear Decommissioning Authority

[11] RIFE 2017 *Radioactivity in Food and the Environment 2016* CEFAS RIFE-22

Chapter 7

Radiation detection and measurement

In chapter 6 we explored some of the ways of undertaking environmental sampling. However, how do we actually measure the radioactivity in the sample? To do this we need to use a radiation detector, whether it is in-built into the environmental monitoring system or the sample is taken to a laboratory where it is measured by a detection system.

Radiation detectors are broadly broken down into two types, passive detectors such as the classic film badge or thermoluminescent materials, or active detectors such as a Geiger counter. The former does not provide a real time measurement of the radiation and tends to be used for measuring external dose to an individual for instance as a personal dosimeter, these are then periodically sent for processing or development to give an indication of the radiation exposure in a period of time (typically monthly or every three months). Meanwhile, active detectors are used to provide real time measurements of radioactivity, dose, or contamination, and as such are most applicable for measuring the activity in a sample and are the focus of this chapter.

Active detectors are based on one of three types of technologies:

- gas based detectors;
- semiconductor detectors;
- scintillation detectors.

In the subsequent sections we will explore each of these technologies in further detail.

7.1 Gas based detectors

Gas based detectors are built on the principle of an ionisation chamber, as seen in figure 7.1.

A moderate voltage is applied between two electrodes, an anode and a cathode, creating an electric field. Ionising radiation that enters the detector may ionise the

doi:10.1088/2053-2563/aafa6dch7

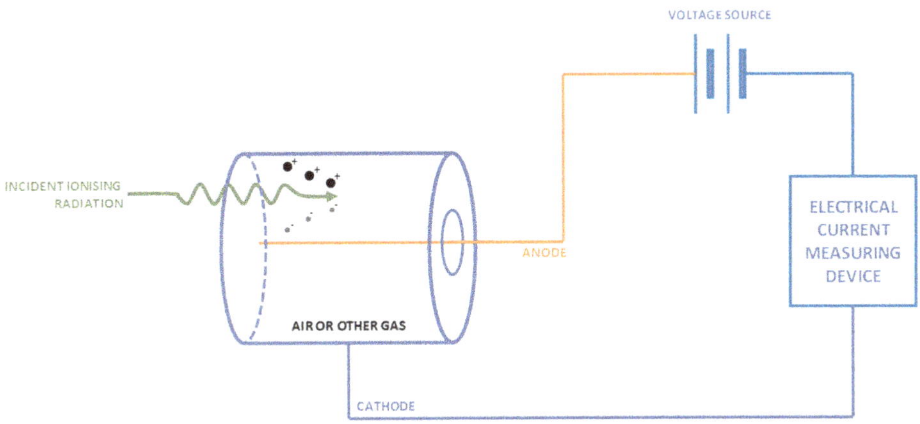

Figure 7.1. Diagrammatic representation of an ionisation chamber.

gas atoms creating an electron–ion pair (as discussed in chapter 1). Due to the electric field the electrons are attracted to the anode and the positive ions to the cathode. This flow of ions creates an electric current which is a measure of the radiation in the gas volume. The electrical current is small and as such an amplifier is used to measure it [1].

The design of the chamber and filling gas used depends on the application it is to be used for. For instance, if the instrument needs to respond to beta particles the chamber walls or front window must be thin to allow the particles to pass through and interact with the gas.

In addition to the design of the chamber and filling gas, the applied voltage to the anode and cathode has an impact in the operation of the detector [2]. This is summarised in figure 7.2 and described further below.

- **Ionisation chamber**. In this region the current measured in the external circuit is equal to the rate of formation of charges in the gas by the incident radiation or intensity of radiation in the gas volume. A common application of this type of detector is as a portable survey meter used to monitor potential personnel exposure to gamma rays.
- **Proportional counter**. In this region the voltage is high enough that the ions pairs produced by the ionising radiation are accelerated by the electric field to a sufficient energy to ionise further gas atoms. Thus, a single photon or ionising particle can produce a pulse of current, a thousand times greater than the pulse produced initially by the radiation interaction. The output is measured as a series of pulses that may be counted by an appropriate means, unlike the ionisation chamber where an average current is measured [3].

  The key importance of this region is that for a certain voltage the ionisation is amplified by a constant amount, such that the number of collected ion pairs is proportional to the initial ionisation. Proportional counters are suitable for more complex radiation measurements such as sorting larger alpha radiation particles from smaller beta particles.

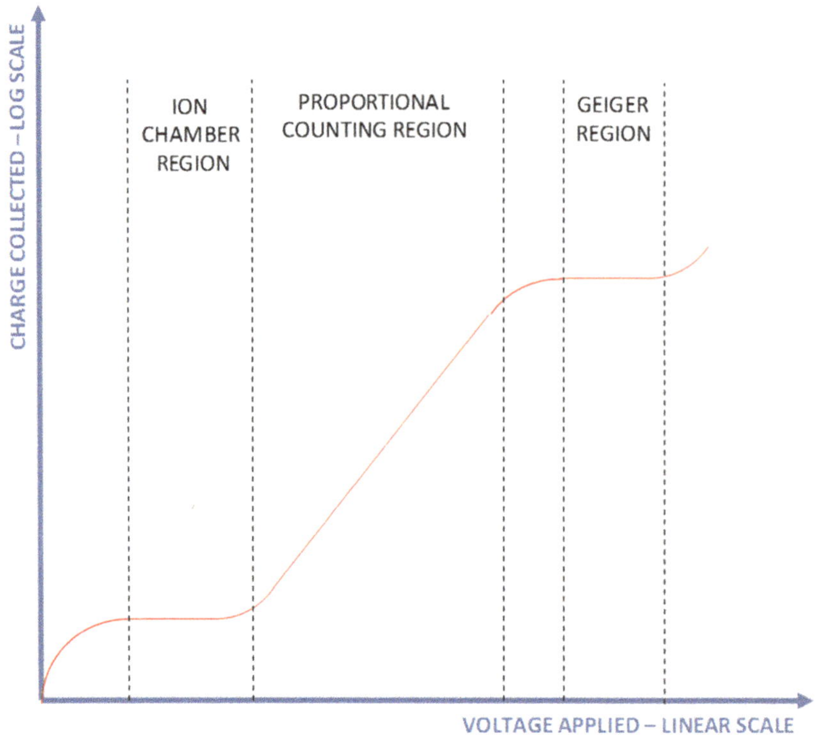

Figure 7.2. Charge collection as a function of an applied voltage in a gas based detector.

- **Geiger–Müller Counter**. In the Geiger–Müller region the amplification of the pulse is so large that a single ionising particle produces an avalanche of ionisation. The size of the pulse is the same regardless of the initiating ionisation. A Geiger–Müller counter is used for counting lightly ionising radiation, such as beta particles and gamma rays, however it cannot normally identify the specific type of radiation emitted or its energy level [4].

7.2 Semiconductor detectors

Semiconductor materials exhibit measurable effects when exposed to ionising radiation. In these materials, electrons exist in defined energy bands, separated by forbidden regions. The highest energy band in which electrons normally exist is the valence band. When ionising radiation is incident on a semiconductor it can transfer its energy to a valence electron and raise it through the forbidden region into the conduction band. The vacancy left by the electron is called a hole and is analogous to a positive ion in a gas detector [5].

When an electric field is applied to the semiconductor, electrons and holes travel to the electrodes, where they result in a pulse that can be measured in an outer circuit. The amount of energy required to create an electron–hole pair is specific to the semiconductor material and temperature and is independent of the energy of the incident radiation [6]. As such, by measuring the number of electron–hole pairs, and the size of the pulse produced, the intensity and energy of the incident radiation can be determined (figure 7.3).

There are two types of semiconductor materials:

- **Intrinsic semiconductors**. These are formed of very high purity materials.
- **Extrinsic semiconductors**. These are formed by adding impurities such as phosphorus and lithium to materials such as germanium (Ge) and silicon. The adding of impurities results in an excess of electrons or hole charge carriers at some energy in the forbidden region, between the conduction and valence band. Thus, creating two sub-categories of extrinsic semiconductors:
    - **n-type**. These materials have an excess of electron charge carriers in a donor band between the valence and conduction band. It is much easier to excite electrons from here to the conduction band than it is from the valence band.
    - **p-type**. These materials have an excess of hole charge carriers in an acceptor band between the valence and conduction band. It is much easier to excite electrons from the valance band to the acceptor band than to the conduction band. The excited electrons leave behind holes in the valence band.

Using these materials two groups of semiconductor detectors can be created:

- **Junction detectors**. An impurity is either diffused into, or oxidised onto a prepared surface of intrinsic material, to change a layer of 'p-type' semiconductor, from or to an 'n-type' semiconductor. When a voltage is applied to the surface barrier, it behaves like a solid version of an ionisation chamber.
- **Bulk conductivity detectors**. These are formed from intrinsic semiconductors of very high bulk resistivity. They also operate like an ionisation chamber, but with a much higher density and greater ionisation per unit absorbed dose.

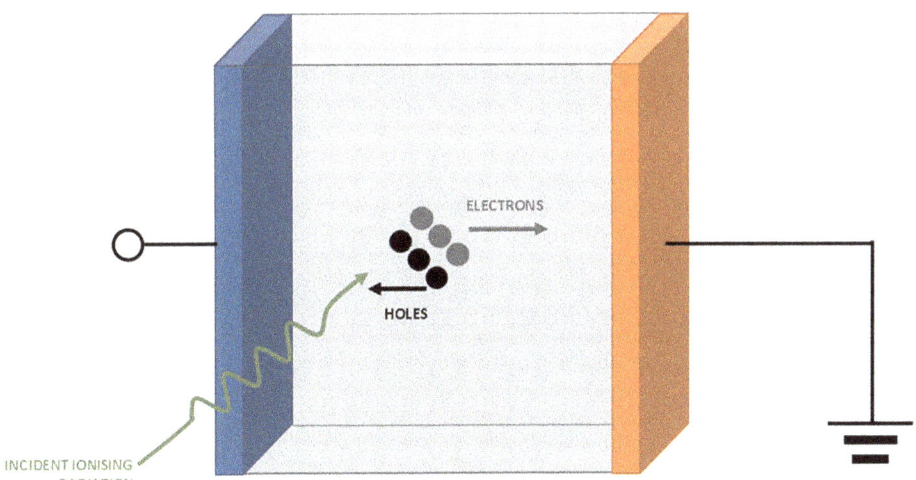

Figure 7.3. Diagrammatic representation of a semiconductor detector.

Semiconductor detectors are used for a range of applications in radiation detection, from alpha and gamma spectroscopy (as discussed in section 7.4), monitoring of radionuclide concentrations in air, such as in environmental monitoring systems, and measuring a range of dose rates.

7.3 Scintillation detectors

Scintillators are usually solids, although liquids or gases can be used. They are based on the principle that they emit light (fluoresce) when ionising radiation interacts with them.

In a similar fashion to a semiconductor the ionising radiation causes the excitation of an electron from the valance band. However, the electron quickly returns from its excited state back to the valance band, typically in the order of about 1 μs. As it returns from its excited state it emits the excess energy in the form of light.

The light is detected by either a photomultiplier tube or photodiode which converts the light into electrical pulses that are subsequently amplified. The size of the pulse is proportional to the energy deposited in the scintillator material by the incident radiation. As such, in a similar fashion to a semiconductor, by measuring the number and size of the pulses produced, the intensity and energy of the incident radiation can be determined (figure 7.4).

Solid scintillators can be used for a wide range of applications in radiation detection such as gamma spectroscopy (as discussed in section 7.4), dose rate measurements and high sensitivity activity measurements.

In addition, liquid scintillators are widely used for the measurement of beta activity in liquid samples, for instance those retrieved from a tritium or carbon-14 bubbler. In this case the sample is mixed with a liquid scintillant and counted using multiple photomultiplier tubes and a coincidence circuit. The coincidence circuit reduces the background of spurious pulses [7, 8].

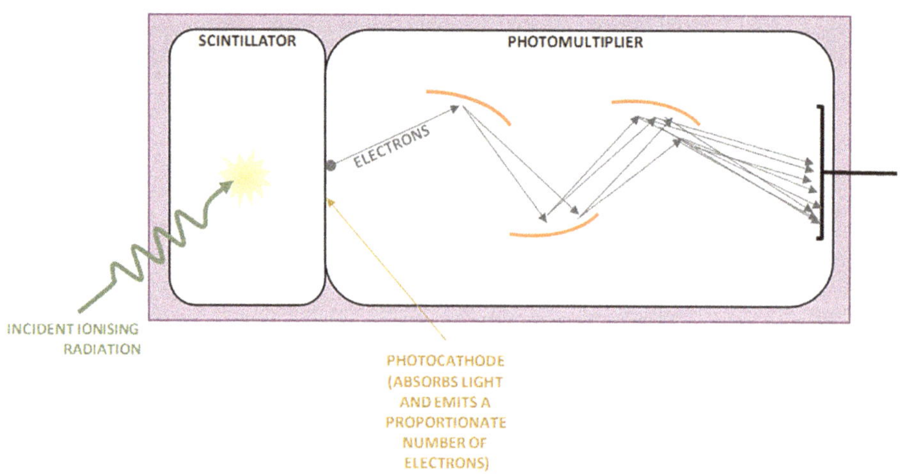

Figure 7.4. Diagrammatic representation of a scintillation detector.

7.4 Counting systems versus spectroscopy

So far, we have covered the technologies used to detect ionising radiation, but how are the signals produced by these systems used to provide us with a useful measurement?

In sections 7.1–7.3 we highlighted that the detector systems either produced an output in the form of an average current, or in pulses. Average current readings are converted directly into a radiation exposure, however, in the case of pulsed outputs, there are two options to extract and analyse the data:

- **Simple counting systems**. In these systems the objective is to record the number of pulses that occur over a measurement time or rate at which the pulses occur. In essence the output is converted into either an indication of radiation intensity, or a reading of activity or dose rate. In these cases, the readings tend to be total readings, for instance total detected activity or dose rate, regardless of type of radiation or radionuclide.
- **Spectroscopy systems**. In many types of detectors such as scintillators, semiconductors, and proportional counters, the amplitude of the pulse is proportional to the energy deposited by the incident radiation. By measuring not only the number of pulses but also their distribution in amplitude it is possible to determine the energy distribution of radiation that is incident on the detector.

  As discussed in chapter 1, radionuclides emit alpha particles, and/or gamma rays at a specific energy. Therefore, by measuring the energy distributions it can be possible to identify the radionuclides from which the ionising radiation incident on the detector originated. This is referred to as alpha spectroscopy and gamma spectroscopy.

A summary of the basic electronics that form part of these two types of systems is presented in figure 7.5.

7.5 Typical applications of detector technologies

As discussed, gas, semiconductor and scintillator detectors can be configured in several ways to allow them to detect different types of radiation and take different types of measurements.

As a guide, to help narrow down the type of detector system suitable for an application, a summary of their typical applications is provided in table 7.1. However, this table should only be considered indicative and consideration should be given to a wide range of criteria when selecting a detector system. Further guidance is provided in section 7.6.

7.6 Criteria for selecting the right detector

As seen, the various types of radiation detectors differ in many aspects. So how do you compare them? And more importantly how do you select the specific one for your application?

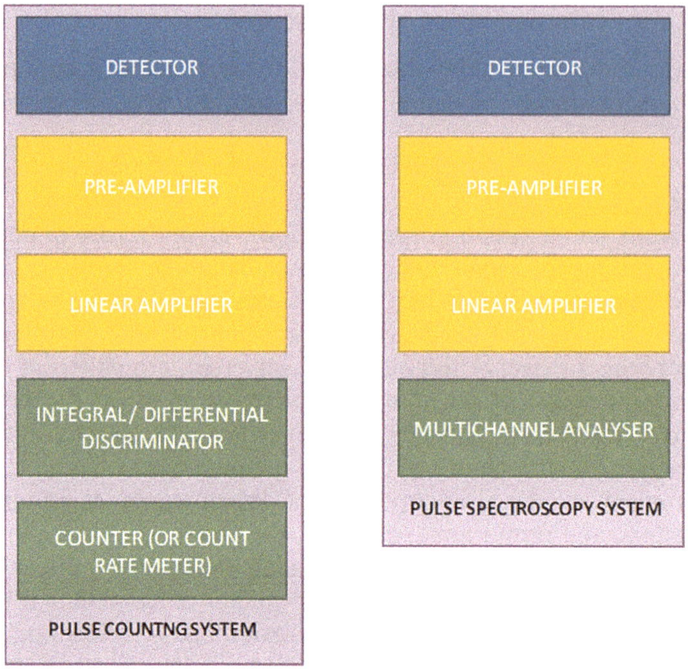

Figure 7.5. Diagrammatic representation of a pulse counting system versus a pulse spectroscopy system.

Table 7.1. Typical application of radiation detector type.

	Type of radiation measured			
Type of detector	Alpha	Beta	Gamma	Used for spectroscopy[a]
Gas filled detector				
– Ionisation chamber	×	×	✓	×
– Proportional counter	✓	✓	✓	(✓)
– Geiger–Müller counter	×	✓	✓	×
Semiconductor detector	✓	×	✓	✓
Scintillation detector	✓	✓	✓	✓

[a] Proportional counters can be used for charged particle spectroscopy, such as differentiating alpha and beta particles ONLY. Semiconductors and scintillation detectors can be used for radionuclide identification *via* gamma ray spectroscopy.

Unfortunately, there is no set process for doing this, and there are country specific standards and regulatory requirements to consider, such as those discussed in chapter 9.

However, as a useful guide a couple of performance-based criteria that are worth considering are detailed below [9, 10]:

- **The sensitivity of the detector**. What type of radiation are you trying to measure? For example, some specific detectors are not suitable for measuring alpha particles because they have a detector covering which the alpha particles cannot penetrate.
- **The type of measurement**. Are you measuring a dose, dose rate, total activity, or specific activities of particular radionuclides? And how precisely does it need to do this? For instance, is it important to be able to distinguish gamma rays at 1.1 and 1.15 MeV?
- **The time response of the detector**. How quickly do you need the detector to respond to a change in the radiation? And for what activity ranges? For example, high or low radiation fields can present challenges in terms of ability to detect small amounts of radiation, or so much radiation the detector electronics cannot process the pulses fast enough.
- **The detector efficiency**. If 1000 gamma rays are incident on the detector, how many do you want to be detected?

In addition to those criteria related to the performance of the detector, it is important to consider those criteria that may impact the design of your new or existing facility, or usability. This includes:

- **The ambient conditions**. What is the response of the detector in the ambient environment where it is to be used? For instance, ambient temperature, humidity, radiofrequencies, magnetic fields etc, can all impact the performance of a detector.
- **The ease of decontamination**. If it is likely the detector will be contaminated taking a measurement, how easy is it to decontaminate? And how frequently will this need to be done? What about the waste produced from the decontamination process?
- **The maintenance and reliability**. How often will the system need to be maintained? How easy will it be to do this? How much will this cost? And how reliable do you need the system to be? What is the life expectancy of the system?
- **The data read-out**. What form do you want the data to be output to? Is it connected to a computer system? Or a read-out directly on the device? What units do you want these to be in?

7.7 Summary

- Radiation detectors are broadly broken down into two types, passive detectors or active detectors.
    - **Passive detectors** do not provide a real time measurement of the radiation and tend to be used for measuring external dose to an individual, for instance as a personal dosimeter.

  - **Active detectors** provide real time measurements of radioactivity, dose, or contamination, and as such are most applicable for measuring the activity in a sample.
- Active detectors are based on one of three types of technology:
  - **Gas based detectors**. Built on the principle of an ionisation chamber. A moderate voltage is applied between two electrodes, creating an electric field. Ionising radiation that enters the detector may ionise the gas atoms creating an electron–ion pair. The flow of ions to the electrodes creates an electric current which is a measure of the radiation in the gas volume. The applied voltage to the anode and cathode has an impact in the operation of the detector:
    - **Ionisation chamber**. In this region the current measured in the external circuit is equal to the rate of formation of charges in the gas by the incident radiation or intensity of radiation in the gas volume.
    - **Proportional counter**. In this region, for a certain voltage the ionisation is amplified by a constant amount, such that the number of collected ion pairs is proportional to the initial ionisation from the incident radiation.
    - **Geiger–Müller counter**. In the Geiger–Müller region the amplification of the pulse is so large, the size of the pulse is the same regardless of the initiating ionisation.
  - **Semiconductor detectors**. Semiconductor materials exhibit measurable effects when exposed to ionising radiation. When ionising radiation is incident on a semiconductor, it can transfer its energy to a valence electron and raise it through the forbidden region into the conduction band. The vacancy left by the electron is called a hole and is analogous to a positive ion in a gas detector. When an electric field is applied to the semiconductor, electrons and holes travel to the electrodes, where they result in a pulse that can be measured in an outer circuit. By measuring the number of electron–hole pairs, and size of the pulse produced, the intensity and energy of the incident radiation can be determined.
  - **Scintillation detectors**. These are based on the principle that they emit light (fluoresce) when ionising radiation interacts with them. The light is detected by either a photomultiplier tube or photodiode which converts the light into electrical pulses that are subsequently amplified. The size of the pulse is proportional to the energy deposited in the scintillator material by the incident radiation.
- Detector systems produce an output in the form of an average current, or pulses. Average current readings are converted directly into a radiation exposure, however, in the case of pulsed outputs, there are two options to extract and analyse the data:
  - **Simple counting systems**. The number of pulses that occur over a measurement time or rate at which the pulses occur is recorded and converted into either an indication of radiation intensity, or a reading of activity or dose rate.

    - **Spectroscopy systems**. Not only the number of pulses but also their distribution in amplitude is recorded, allowing the energy distribution of radiation that is incident on the detector to be determined. By measuring the energy distributions it can be possible to identify the radionuclides from which the ionising radiation incident on the detector originated. This is referred to as alpha spectroscopy and gamma spectroscopy.
- There is no set process for selecting the right detector system for an application. In addition, there are country specific standards and regulatory requirements to consider. However, regardless of the country there are a number of criteria that should be considered when selecting a detector system:
    - Performance criteria:
        - detector sensitivity;
        - types of measurements required;
        - time response of the detector;
        - detector efficiency.
    - Usability and facility integration criteria:
        - ambient environmental conditions;
        - ease of decontamination;
        - maintenance and reliability;
        - data read-out.

References

[1] Martin A, Harbison S, Beach K and Cole P 2012 *An Introduction to Radiation Protection* 6th edn (Boca Raton, FL: CRC Press)

[2] Knoll G F 2000 *Radiation Detection and Measurement* 3rd edn (New York: Wiley)

[3] IAEA 2004 Workplace monitoring for radiation and contamination *Practical Radiation Technical Manual IAEA PRTM-1 (Rev. 1)*

[4] Knoll G F Radiation measurement technology *Encyclopaedia Britannica* https://britannica.com/technology/radiation-measurement/Passive-detectors [accessed 12 August 2018]

[5] Ridha A A 2013 *Determination of Radionuclides Concentrations in Construction Materials Used in Iraq* (Baghdad, Iraq: University of Al-Mustansiriyah College of Science)

[6] Bevan A *Semiconductor Detectors Pedagogical Review of Aspects of Organic and Inorganic Devices* (London: Queens Mary University of London) https://pprc.qmul.ac.uk/~bevan/detectors/SemiconductorDetectors.pdf [accessed 12 August 2018]

[7] IAEA 2004 *Management of waste containing tritium and carbon-14Technical Reports Series No. 421* (Austria: IAEA)

[8] IAEA 1991 *Safe handling of tritium review of data and experienceTechnical Reports Series No. 324* (Austria: IAEA)

[9] ENSREG *Environmental Monitoring* http://ensreg.eu/nuclear-safety/environmental-monitoring [accessed 15 July 2018]

[10] IAEA 2005 *Environmental and source monitoring for purposed of radiation protectionIAEA Safety Standards Series No. RS-G-1.8* (Austria: IAEA)

Chapter 8

International legislation and standards

So far, we have focused on the technical aspects of radioactive discharges, such as what is radiation, the sources of radioactive discharges, how it moves in the environment, the impacts on human health and how do we measure these impacts.

In chapters 8 and 9 we will explore how radioactive discharges are controlled from a legal point of view.

A wide variety of laws have been developed to protect the environment from a range of pollutants (substances, energy, vibration and noise) that, when introduced to the environment, have harmful effects. The early laws focused on the protection of rights associated with property ownership, which indirectly protected the environment, whilst more recent laws protect the environment itself.

Some domestic environmental laws (national laws) originate from international conventions and agreements. For instance, in the UK 90% of UK environmental laws originate from the EU. International treaties seek to regulate issues as diverse as climate change, international waste shipments, access to environmental information and transboundary impacts such as radioactive discharges.

In the following chapter we will explore the origin of the principles that govern the protection of humans and the environment from radioactive discharges and look at how these have been incorporated into international conventions and agreements. Chapter 9 will then look at the incorporation of these conventions and agreements into UK domestic law.

8.1 International commission on radiological protection (ICRP) recommendations and standards

The principles for protecting humans and the environment from the harmful effects of radiation originate from the international commission on radiological protection (ICRP) [1].

doi:10.1088/2053-2563/aafa6dch8

ICRP is an independent, international organisation with members from approximately 30 countries across six continents. These members represent the leading scientists and policy makers in the field of radiological protection.

Established in 1928 the ICRP has developed and maintained the 'International System of Radiological Protection' used as the international common basis for radiation protection standards, legislation, and practice. The system is based on the current understanding of the science of radiation exposure and effects and value judgements taking account of societal expectations, ethics and experience.

In addition to the 'International System of Radiological Protection' the ICRP publishes reports on all aspects of radiological protection, including dose conversion coefficients as discussed in chapter 5, radionuclide data, models for measuring dose effects and industry specific guidance.

8.1.1 ICRP publication 103—the 2007 recommendations of the ICRP

The recommendations of the ICRP form the basis of the 'International System of Radiological Protection' discussed earlier.

The earlier 1990 recommendations were formed using a process-based system with different principles for practices and interventions [2], where:

- Practice is a human activity that increases radiation exposure over and above that incurred from background or increases the likelihood of incurring an exposure.
- Intervention is a human activity that seeks to reduce existing radiation exposures or reduce existing likelihood of exposure which is not part of a practice.

The aim of the 1990 recommendations was to establish a method for evaluating the risk from exposures to humans and set limits based on maximum tolerable individual risk.

The 2007 recommendations underpin the current 'International System of Radiological Protection'. The majority of the changes were minor in nature. However, a couple of key updates included:

- Updated weighting factors and detriment (see chapter 5).
- The importance of radiation protection in medicine and the environment highlighted.
- Focus on the exposure situation rather than a process-based approach to practice or intervention.
- The introduction of a source-related constraint extended to all situations.

The primary aim of the new recommendations is 'to contribute to an appropriate level of protection for people and the environment without unduly limiting the desirable human activities that may be associated with radiation exposure'.

Three exposure-based situations are covered under the 2007 recommendations:

- **Planned exposures**. Everyday situations involving the planned operation of practices (such as planned discharges of radioactivity).
- **Emergency exposures**. Unexpected situations that occur during operation of a practice requiring urgent action.

- **Existing exposures**. Situations that already exist when a decision on control has to be taken, including natural background radiation and residue from past practices.

Within each exposure situation the following types of exposure are included:
- **Occupational exposure**. incurred by workers in the course of their work.
- **Public exposure**. incurred by members of the public, excluding occupational or medical exposure of patients.
- **Medical exposure**. incurred by patients as part of their own medical or dental diagnosis or treatment; volunteers helping in the support and comfort of patients; and biomedical research volunteers.

The 'International System of Radiological Protection' applies three protection principles to these exposure situations and types:
- **Justification**. Any decision that alters a radiation exposure situation should do more good than harm.
- **Optimisation of protection**. The likelihood of incurring exposures, the number of people exposed, and the magnitude of their individual doses should all be kept As Low As Reasonably Achievable (ALARA), taking account of economic and societal factors.

  In essence when optimising the risk associated with the exposure to ionising radiation, this should not simply involve minimisation of the exposure at all cost, but should be a balance of the health effect reduction, costs involved and other relevant factors such as waste generation, ease of maintenance and technological feasibility.

  To avoid severely inequitable outcomes of the optimisation procedure, there are restrictions on the doses or risks to individuals from a source, these are:
  - **Constraints**. Prospective, source-related restriction on the dose to the most highly exposed individuals from the planned operation of any controlled source.
  - **Reference level**. In emergency or existing controllable exposure situations, this represents the levels or dose or risk, above which it is judged to be inappropriate to plan to allow exposures to occur and below which optimisation of protection should be implemented.

  Typical bands for effective dose constraints and reference levels are provided below:
  - **20–100 mSv yr^{-1}**. Reference level for evacuation in an emergency. Occupational exposures in a rescue operation.
  - **1–20 mSv yr^{-1}**. Constraints set for occupational exposure and for comforters and carers. Reference level for radon.
  - **< 1 mSv yr^{-1}**. Constraints set for public exposure in planned situations.
- **Dose limitation**. The total dose to any individual from all regulated sources in planned exposure situations, other than medical exposure of patients, should not exceed the appropriate limits recommended by the ICRP.

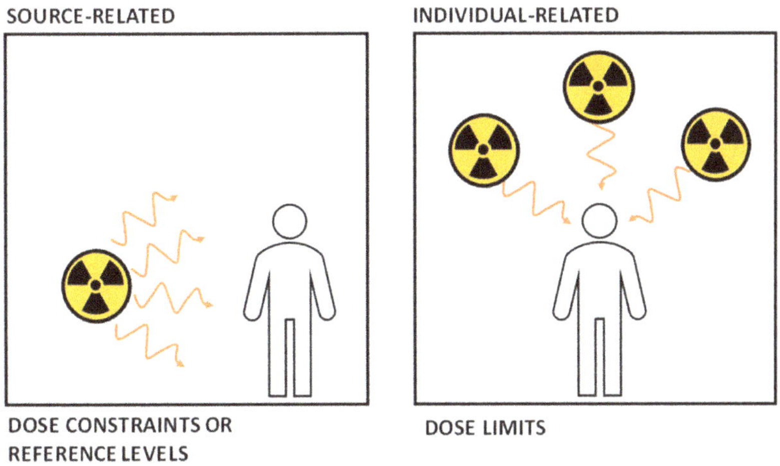

Figure 8.1. Diagrammatic representation of source related and individual related protection.

Of the three principles, the first two are source related for all situations and the last is only related to planned exposure situations, namely:

- The individual is protected from a single source in planned, emergency and existing situations by the dose constraint or reference levels.
- The individual is protected from all regulated sources in planned situations by the dose limits.

This is summarised in figure 8.1.

The recommendations emphasise that optimisation of protection is at the heart of a successful radiological protection programme. In order to successfully embed the optimisation principle into a radiological protection programme commitment is required from all those involved. For instance, imagine the design of a new nuclear power station: commitment of management to reduce radiological exposures to the workforce and public under both planned and emergency situation is key, but so also is that of the workforce and the regulatory bodies enforcing the law and regulatory system.

To do this, ensuring the correct culture within the organisation is key and should be enforced through training and good management practices. This culture is referred to as 'safety culture' and is defined as 'the assembly of characteristics and attitudes in organisations and individuals which establishes that, as an overriding priority, protection and safety issues receive the attention warranted by their significance'.

8.2 International conventions, standards and directives

There are a number of international conventions, standards and directives that govern the protection of humans and the environment from radioactive discharges. The main one of these is the International Atomic Energy Agency Basic Safety Standards. This is discussed in section 8.2.1.

8.2.1 International Atomic Energy Agency Basic Safety Standards

The International Atomic Energy Agency (IAEA) is the international centre for cooperation in the nuclear field, promoting the safe, secure, and peaceful use of nuclear technologies. Through its statute it is specifically authorised to establish standards of safety for the protection of health and minimisation of danger to life. The IAEA develops such standards on the basis of an open and transparent process for gathering, integrating and sharing the knowledge and experience gained from the use of technologies and from the application of the safety standards themselves. Where applicable these are adopted by member states, in particular by regulatory bodies and other relevant national authorities [3].

Of particular relevance to radioactive discharges is the IAEA International Basic Safety Standards on Radiation Protection and Safety of Radiation Sources (BSS).

The BSS details the requirements for the protection of people and the environment from harmful effects of ionising radiation and for the safety of radiation sources. It considers all radiation exposure situations rather than those specifically related to nuclear facilities.

The basis of the standards takes into account the findings and the recommendations of the ICRP (ICRP Publication 103) along with those from other organisations such as the United Nations Scientific Committee on the Effects of Atomic Radiation (UNSCEAR).

This includes the principles on:

- radiation effects;
- exposure situations and types, and
- the 'International System of Radiological Protection'.

There are 52 requirements in total within the BSS, broken down into four categories:

- general requirements for protection and safety;
- planned exposure situations;
- emergency exposure situations;
- existing exposure situations.

Of particular relevance to the radioactive discharges and protection of the environment are:

- **Requirement 29: Responsibilities of the government and the regulatory body specific to public exposure**. This sets out the requirement for the national government to establish a regulatory body to enforce the requirements for optimisation and dose limitation of public exposures.
- **Requirement 30: Responsibilities of relevant parties specific to public exposure**. This sets out the requirement for all parties involved on the use, storage or disposal of radioactive material to apply the 'international system of radiological protection' to protect members of the public against radiation exposure.
- **Requirement 31: Radioactive waste and discharges**. This sets out the requirement for all relevant parties to ensure that any radioactive waste and

discharges to the environment are managed in accordance with an authorisation granted by the regulatory body. The authorisation will include operational limits and conditions relating to public exposure, which have been set based on the prospective and retrospective assessment of the environmental impacts. This is discussed further in chapter 9.
- **Requirement 32: Monitoring and reporting**. This sets out the requirement for regulatory bodies to ensure that a programme is in place for source monitoring and environmental monitoring and that the results from the monitoring are recorded and made available on request of the IAEA.

8.2.2 International conventions relevant to radioactive discharges and radioactive waste

In addition to the requirements laid out by the IAEA, there are a number of conventions relevant to radioactive discharges and radioactive waste. These are summarised below:

- **The joint convention on the safety of spent fuel management and on the safety of radioactive waste management 1998 (The IAEA joint convention)**. The convention applies to both spent fuel and radioactive waste [4]. It also applies to planned and controlled releases into the natural environment of liquid and gaseous radioactive materials from regulated nuclear facilities. However, it does not apply to NORM (enhanced naturally occurring radioactive material). It also excludes military radioactive waste. The convention aims to protect 'individuals, society and the environment' from harmful effects of ionising radiation through the enhancement of national measures and international cooperation, including where appropriate safety related technical cooperation.
- **Convention on the prevention of marine pollution by dumping of wastes and other matter (The London convention 1972 and the protocol to the London convention 1996)**. The 1972 London convention and the 1996 protocol to the convention were adopted, under the auspices of the International Maritime Organisation (IMO), to ban the dumping of all types of waste, including radioactive waste and other radioactive matter at sea (all marine waters other than the internal waters of states) [5]. Protection of the marine environment is the aim, and hence long based discharges of radioactive waste, and discharge of radioactivity into the air does not fall under the scope of these legal instruments.
- **The convention for the protection of the marine environment of the North-East Atlantic (The OSPAR convention 1992)**. This applies to the marine environment only [6]. Geographically, only the marine environment of the North-East Atlantic falls under its scope. It has a wide scope covering hazardous and radioactive substances in the sea. This includes both anthropogenic (man-made) radioactive substances and NORM (enhanced naturally occurring radioactive materials). It stipulates that all man-made radioactive discharges must reach zero levels. It applies to both land-based discharges and dumping of all radioactive substances.

8.3 The EURATOM treaty and directives

The Euratom treaty was signed on 25 March 1957 establishing the European Atomic Energy Community. The treaty provides a regulatory framework that governs both the use of nuclear materials, such as in the civil nuclear industry, and the wider use of radioactive substances in non-nuclear sectors [7].

Of particular relevance to radioactive discharges is chapter 3 of Title II of the EURATOM treaty 'Health and Safety', which contains a number of articles [8, 9]. These include:

- **Articles 30–33**. These articles cover the establishment of the European basic safety standards covering the protection of health of workers and the general public against the dangers arising from ionising radiations, and requires the member states of the European atomic energy community to incorporate the standards into their domestic legislation. The European basic safety standards are broadly in line with the IAEA BSS and as such by complying with the European basic safety standards all members of the European Atomic Energy Community also meet the requirements of the IAEA BSS.
- **Articles 35 and 36**. These articles establish the requirement for the member states of the European Atomic Energy Community to undertake continuous monitoring of the levels of radioactivity in the air, water and soil and provide the European Commission with periodic reports on the data collected.
- **Article 37**. This article requires the submission of 'general data' relating to any plan for the disposal of radioactive waste, to determine if there is any potential to result in radioactive contamination of the water, soil or airspace of another member state.

8.4 Summary

- Early environmental laws focused on the protection of rights associated with property ownership, which indirectly protected the environment, whilst more recent laws protect the environment itself.
- International treaties seek to regulate issues as diverse as climate change, international waste shipments, access to environmental information and transboundary impacts such as radioactive discharges.
- The principles for protection of humans and the environment from the harmful effects of radiation originate from the International Commission on Radiological Protection (ICRP).
- The recommendations of the ICRP (ICRP Publication 103) form the basis of the 'International System of Radiological Protection'. It covers planned exposure, emergency exposure and existing exposures and is based on three protection principles:
    - **Justification**. Any decision that alters a radiation exposure situation should do more good than harm.
    - **Optimisation of protection**. The likelihood of incurring exposures, the number of people exposed, and the magnitude of their individual doses

  should all be kept As Low As Reasonably Achievable (ALARA), taking account of economic and societal factors.
  - **Dose limitation**. The total dose to any individual from all regulated sources in planned exposure situations, other than medical exposure of patients, should not exceed the appropriate limits recommended by the ICRP.

  Of the three principles the first two are source related for all situations and the last is only related to planned exposure situations.
- The ICRP 2007 recommendations emphasise that optimisation of protection is at the heart of a successful radiological protection programme. 'Safety culture' plays a key role in ensuring this.
- There are a number of international conventions, standards and directives that govern the protection of humans and the environment from radioactive discharges. These include:
  - **International Atomic Energy Agency Basic Safety Standards**. The BSS details the requirements for the protection of people and the environment from harmful effects of ionising radiation and for the safety of radiation sources. It considers all radiation exposure situations. The basis of the standards takes account the findings of the recommendations of the ICRP (ICRP Publication 103) along with those from other organisations such as the United Nations Scientific Committee on the Effects of Atomic Radiation (UNSCEAR).
  - **The joint convention on the safety of spent fuel management and on the safety of radioactive waste management 1998 (The IAEA joint convention)**. The convention applies to both spent fuel and radioactive waste. The convention aims to protect 'individuals, society and the environment' from harmful effects of ionising radiation through the enhancement of national measures and international cooperation, including where appropriate safety related technical cooperation.
  - **Convention on the prevention of marine pollution by dumping of wastes and other matter (The London convention 1972 and the protocol to the London convention 1996)**. The 1972 London convention and the 1996 protocol to the convention were adopted, under the auspices of the International Maritime Organisation (IMO), to ban the dumping of all types of waste, including radioactive waste and other radioactive matter at sea (all marine waters other than the internal waters of states).
  - **The convention for the protection of the marine environment of the North-East Atlantic (The OSPAR convention 1992)**. It applies to the marine environment only. Geographically, only the marine environment of the North-East Atlantic falls under its scope. It stipulates that all man-made radioactive discharges must reach zero levels. It applies to both land-based discharges and dumping of all radioactive substances.
  - The EURATOM treaty established the European Atomic Energy Community. The treaty provides a regulatory framework that governs both the use of nuclear materials, such as in the civil nuclear industry,

and wider use of radioactive substances in non-nuclear sectors. Of particular relevance to radioactive discharges are the following articles:

- **Articles 30–33**. These articles cover the establishment of the European basic safety standards covering the protection of the health of workers and the general public against the dangers arising from ionising radiations and requires the member states of the European Atomic Energy Community to incorporate the standards into their domestic legislation. The European basic safety standards are broadly in line with the IAEA BSS.
- **Articles 35 and 36**. These articles establish the requirement for the member states of the European Atomic Energy Community to undertake continuous monitoring of the levels of radioactivity in the air, water and soil and provide the European Commission with periodic reports on the data collected.
- **Article 37**. This article requires the submission of 'general data' relating to any plan for the disposal of radioactive waste, to determine if there is any potential to result in radioactive contamination of the water, soil or airspace of another member state.

References

[1] Martin A, Harbison S, Beach K and Cole P 2012 *An Introduction to Radiation Protection* 6th edn (Boca Raton, FL: CRC Press)

[2] ICRP 2007 The 2007 recommendations of the international commission on radiological protection ICRP publication 103 *Ann. ICRP* **37**

[3] IAEA 2014 *Radiation Protection and Safety of Radiation Sources: International Basic Safety Standards General Safety Requirements Part 3*

[4] IAEA 1997 *The Joint Convention on the Safety of Spent Fuel Management and on the Safety of Radioactive Waste Management* INFCIRC/546

[5] *1996 Protocol to the Convention on the Prevention of Marine Pollution by Dumping of Wastes and Other Matter 1972* http://imo.org/en/OurWork/Environment/LCLP/Documents/PROTOCOLAmended2006.pdf [accessed 2 September 2018]

[6] *OSPAR Convention* https://ospar.org/convention [accessed 2 September 2018]

[7] European Union 2016 *The Euratom Treaty Consolidated Version* http://consilium.europa.eu/media/29775/qc0115106enn.pdf [accessed 2 September 2018]

[8] European Union 2013 *Council Directive 2013/59/Euratom of 5 December 2013 laying down basic safety standards for protection against the dangers arising from exposure to ionising radiation, and repealing Directives 89/618/Euratom, 90/641/Euratom, 96/29/Euratom, 97/43/Euratom and 2003/122/Euratom*

[9] Barrett J, Broughton J, Bryant P, Clark S, Perks C, Thurston J and Webbe-Wood D 2018 *Exit from the EU and the EURATOM Treaty: The Implications for Radiation Protection in the UK*

IOP Publishing

Chapter 9

UK regulation and guidance

In chapter 8 we covered international conventions, standards and directives for controlling radioactive discharges and their transboundary impacts. The following chapter looks at how these are implemented within domestic (national) legislation within the UK.

As discussed earlier, the ICRP recommendations [1] provide the basis of the IAEA basic safety standards [2], and the EURATOM basic safety standards directive (BSSD) [3, 4]. As a member state under the EURATOM treaty it is mandated that the EURATOM BSSD is incorporated into UK domestic legislation.

The UK's domestic legislation for radiation protection (including the control of radioactive discharges) has recently been updated to incorporate the latest EURATOM BSSD and ICRP recommendations (Publication 103) [5]. A summary of the current UK regulatory framework for radiation protection is summarised in figure 9.1.

As seen above, the UK's regulatory framework is divided into a number of parts. These are based on what the radiation is used for, type of facility using/discharging radioactive material, and whether the discharges are planned or unplanned.

A summary of the key regulations controlling radioactive discharges is provided below:

- **Environmental Permitting Regulations 2016 (as amended)**. Regulates the planned discharge of radioactivity from nuclear and non-nuclear establishments across England and Wales. Additionally, it covers the storage and use of radioactive materials on non-nuclear establishments.
- **Environmental Authorisations (Scotland) Regulations 2018**. Equivalent to the environmental permitting regulations 2016 in Scotland.
- **Nuclear Installations Act 1965**. Regulates the activities undertaken on a nuclear establishment including the storage and use of radioactive materials and discharges of radioactive material under unplanned events.

doi:10.1088/2053-2563/aafa6dch9

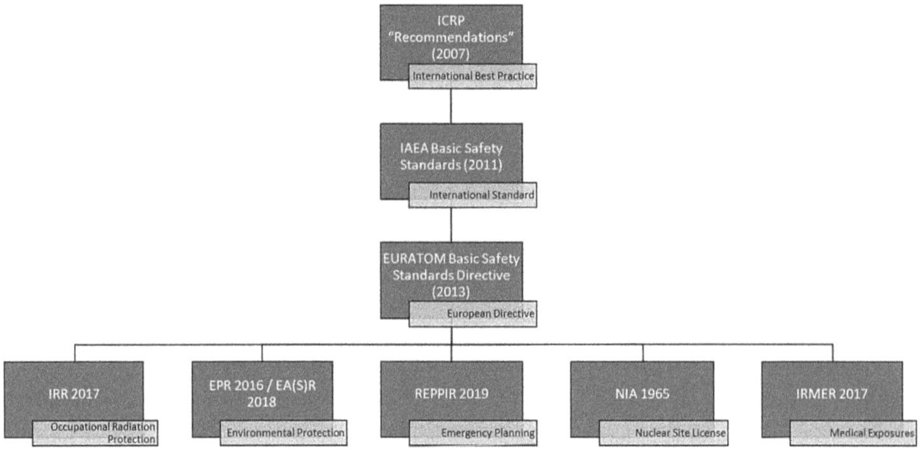

Figure 9.1. Diagrammatic representation of flow-down of the ICRP recommendations into the UK regulatory regime.

- **Ionising radiations regulations 2017**. Regulates the use of radioactive materials and artificial sources to ensure that adequate protection is in place for individuals working with radiation and those that could be impacted by its use (e.g. visitors, members of the public).
- **Radiation (emergency preparedness and public information) regulations 2019**. Establishes a framework to ensure that members of the public are properly informed and prepared in advance about what to do in a radiation emergency and provided with information should an event occur.

In the following sections we will explore these regulations in further detail to see how they ensure a holistic approach to protecting people from the potential effects of radioactive discharges. We will also look at the potential impacts of the UK exit from the European Union and the EURATOM Treaty (BREXIT) on the future of the regulation of radioactive discharges in the UK.

9.1 Environmental permitting regulations

The environmental permitting regulations [6] requires operators of certain facilities that could harm the environment or human health to obtain a 'permit' or 'permits' to allow them to undertake certain activities associated with the use of certain substances and the generation and disposal of waste.

The 'permit' provides a means for ongoing supervision by the regulators to ensure compliance with the conditions of the permit(s) and certain environmental targets, encourage best practice in the operation of facilities, and to fully implement European Legislation [7].

The permitting regime also requires operators who do not require a permit to register as exempt.

In England the regulator who enforces this regime is the Environment Agency and in Wales it is Natural Resources Wales.

Schedule 23 of the regulations relate specifically to the control of radioactive substances, which includes radioactive materials, solid waste and discharges [8]. It requires all operators who will be generating, treating and disposing of radioactive waste to obtain a Radioactive Substances Regulation (RSR) Permit, regardless of whether they are a nuclear establishment or not. It also requires non-nuclear establishments to obtain a permit for the use and storage of radioactive materials such as radiography sources or check sources used for calibrating radiation detectors.

The schedule also defines radioactivity limits for solid materials and waste below which the material/waste does not fall under the legal definition of radioactive, but above which it does. These wastes/materials are referred to as 'out of scope'.

As part of the RSR permit application the operator is required to describe the scope of the activities involving the use of radioactive material; in particular, how any radioactive waste is generated, treated and ultimately disposed of is appropriately optimised to ensure the end impact to the environment and the public is ALARA.

In order to do this the operator must describe in a structured manner how it applies the principle of 'best available techniques' or BAT for short [9], where:

- 'techniques' includes both the technology used and the way in which installation is designed, constructed, maintained operated and decommissioned
- 'best' means the most effective in achieving a high level of protection of the public from exposure to ionising radiation, and
- 'available' requires consideration of whether the techniques have been developed on a scale which allows implementation and whether the techniques are economically and technically viable, taking account of the both the benefits and detriments.

As part of their permit application, the operator must also propose limits to any discharges of radioactivity, assess the potential radiological impacts of these discharges to humans and the environment (including the flora and fauna), and describe how it will monitor these discharges and the environmental impacts [10]. The assessment of the potential impacts must cover all the exposure pathways discussed in chapter 4 of this book. In addition, the operator must describe their management arrangements for how it will ensure compliance with the permit should it be awarded.

The regulator will assess the permit application, which will include doing their own independent assessment of the radiological impacts along with the other information provided in the application. As part of this assessment the radiological impacts to the public are compared against the criteria presented in table 9.1.

The dose limit is a legal limit representing the total dose to any individual from all regulated sources in planned exposure situations that must not be exceeded. The site and source constraint are recommended upper bounds that members of the public may receive from the planned operation of a controlled source. The site constraint represents the constraints from all sources of radioactive discharges from a single site, for instance where a site has more than one nuclear power station present, and

Table 9.1. Criteria used in the assessment of radiological impacts.

Effective dose criteria	Quantity (mSv yr^{-1})
Dose limit	1.00
Site constraint	0.50
Source constraint	0.30
No regulatory concern	$\leqslant$ 0.01

the source constraint represents the upper bound from a single source, such as a single nuclear power station.

Lastly, the threshold 'no regulatory concern' represents a value below which the regulators should not seek to secure further reductions in the exposure of members of the public, subject to the continued application of BAT.

Depending on the outcome of their assessment, the regulator will decide whether or not to award the permit and what limits and conditions should apply.

Only once the permit is awarded and any relevant conditions met will the operator be allowed to begin undertaking the activities associated with their permit.

Throughout the operation of the facility and validity of the permit, the operator will be required to meet the conditions of the permit. The exact conditions will vary depending on the facility but will include a requirement to continue to apply the principle of BAT, a requirement to monitor any waste arisings and discharges and periodically report these to the regulator and notify the regulator should they breach any limits or conditions of the permit.

Any changes in the activities involving the use of radioactive materials, or generation, treatment or disposal of radioactive waste, for instance as the facility moves towards decommissioning will also need to be communicated to the regulator. Depending on the significance of the change a permit variation may be required.

When all activities involving the management of radioactive waste have ceased the operator is required to surrender their environmental permits.

9.2 Environmental Authorisations (Scotland) Regulations

The Environmental Authorisations (Scotland) Regulations [11] are broadly equivalent to the Environmental Permitting Regulations but relate to Scotland.

The main difference in the regulations is that rather than awarding a permit, the regulator, which is the Scottish Environment Protection Agency, awards an authorisation.

Much like the environmental permitting regulations, the Environmental Authorisations (Scotland) Regulations have a schedule (Schedule 8) specifically relating to radioactive substances. This also covers use of radioactive materials, and generation and disposal of solid radioactive waste and discharges.

9.3 Nuclear Installations Act

The Nuclear Installations Act [12] is the main act of UK parliament that relates to nuclear power plants and other nuclear establishments across the UK. It established the UK's nuclear regulator, the Office for Nuclear Regulation (ONR), and put absolute liability on the operator of the nuclear establishment.

This means that a company, irrespective of negligence, is liable for personal injury to any person and/or damage to third party property, associated with the use of fissile material or a by-product of the nuclear reaction.

In order to assure the safety of nuclear establishments in the UK, the ONR operates a system of regulatory control based on a robust licensing process, similar to the environmental permitting regime.

Prior to the construction and operation of a nuclear establishment a corporate body is granted a Nuclear Site Licence (NSL) to use a site for specified activities.

The NSL is a legal document, issued for the full life-cycle of the facility and defines a set of 36 standard conditions, covering design, construction, operation and decommissioning [13].

These conditions require licensees to implement adequate arrangements to ensure compliance. Of the 36 conditions, a number are directly applicable to the control of radioactive discharges. In particular, the licensee is required to demonstrate to the ONR throughout the various stages of the design, construction, commissioning, operation and decommissioning of the establishment, how all risks associated with the establishment are reduced to so far as is reasonably practicable (SFAIRP, the UK equivalent to the ALARA principle). This includes identification of those scenarios resulting in an accidental release of radioactivity into the environment, and how the design and operation of the establishment adequately prevents or mitigates against the impacts to the environment and the public. This is documented in the form of a nuclear safety case which is produced and updated at various stages of the design, construction and commissioning of the establishment, and periodically maintained during operation.

In addition, the NSL requires that all changes to the design or operation of the establishment must first be approved by the ONR, and when the establishment is finally closed down the nuclear decommissioning processes must be formally approved by the ONR.

It should be noted that in addition to regulating nuclear safety, ONR independently regulates security at the 37 nuclear licensed sites in the UK to prevent the theft or sabotage of nuclear or other radioactive materials.

This includes the movement of flasks carrying spent nuclear fuel from operating and decommissioning nuclear reactors, radio-pharmaceuticals needed for hospitals, sealed radioactive sources needed in the construction industry and, for instance, in the non-destructive testing of North Sea oil rigs.

9.4 Ionising radiations regulations

The Ionising Radiations Regulations [14] form part of the Health and Safety at Work Act and require steps to be taken to ensure that radiation exposures to workers are restricted so far as is reasonably practicable (SFAIRP) [15].

Table 9.2. Legal dose limits.

	Dose limit (mSv per annum)		
	Adult workers	Trainees (under 18 years)	Members of the public
Whole body	20	6	1
Lens of eye	20	15	15
Skin (per cm)	500	150	50
Hands, forearms, feet and ankles	500	150	50

This is achieved by placing duties on employers to protect employees and other persons against ionising radiation, arising from their work with radioactive substances and other sources of ionising radiation. It additionally places certain duties on employees.

The ionising radiations regulations are enforced by the health and safety executive (HSE) for non-nuclear establishments and office for nuclear regulation (ONR) for nuclear licensed sites.

Although primarily focused on the protection of the worker it also covers individuals that could be impacted by the employers use of radiation such as visitors to the facility or members of the public.

One of the key requirements of the regulations is to conduct a radiation risk assessment, which includes identifying who can be harmed and how and requires measures to be put in place to prevent, control or restrict any radiation exposures SFAIRP.

For nuclear establishments this requirement has a strong interface with the nuclear safety case. With the radiation risk assessment covering day-to-day operations involving the use of radiation, such as radiation surveys, and the nuclear safety cases covering the design and operation of the facility.

In the case of non-nuclear facilities this risk assessment covers both day-to-day operations and the design of the establishment. For instance, the use of radiopharmaceuticals or x-ray equipment in a hospital. This includes identification of those scenarios resulting in an accidental release of radioactivity into the environment and how they are prevented, controlled or mitigated.

The ionising radiations regulations also set out the legal dose limits for workers and members of the public in line with the IAEA and EURATOM basic safety standards and ICRP publication 103, a summary of which is presented table 9.2.

These limits refer to the total dose to any individual from all regulated sources in planned exposure situations, other than medical exposure of patients and background radiation and should not be exceeded.

9.5 Radiation (emergency preparedness and public information) regulations

Although highly unlikely, the radiation (emergency preparedness and public information) regulations [16] or REPPIR for short, recognises the need to have a plan in place in the of event of an accident or incident resulting in a large radiological release to the environment.

For nuclear establishments there is a strong link between REPPIR and NSL, with the NSL requiring the operator to have adequate arrangements for dealing with any accident or emergency arising on their site and their effects. This includes incidents on the site, with only a small or no radiological release, such as a fire or security incident and those with a large radiological release [17].

REPPIR on the other hand relates specifically to a 'off-site emergency', namely where there is a large release of radioactivity into the environment which results, or is likely to result, in the need to consider urgent countermeasures to protect the public outside the site from the radiological hazard.

The regulations are enforced by the health and safety executive and establish a framework of roles and responsibilities for local responders including the duty to assess the risk of an emergency occurring and to maintain plans for the purposes of responding to an emergency.

The area where members of the public would be likely to be affected by a radiation emergency is called the 'REPPIR off-site emergency planning area' or 'detailed emergency planning zone' (DEPZ).

This area is normally also defined as the 'REPPIR prior information area' within which information is required to be provided by the operators to any identifiable population group so they can be prepared should an incident or event occur.

The local county council will produce and maintain the off-site plan for the DEPZ, usually in collaboration with the operator. Within this area a zone will be identified within which urgent countermeasures (sheltering and potassium iodate tablets) could be justified in the event of the worst radiation emergency.

The off-site plan may also additionally define a food restriction area, within which restrictions on food might be applied for a reasonably foreseeable accident. It also may cover extendibility arrangements.

9.6 Impact of BREXIT

On 29 March 2017 the UK Prime Minister informed the European Union (EU) of the UK's intention to withdraw from both the European Union and the European Atomic Energy Community (EURATOM). Ultimately this will mean that the UK will no longer be part of EURATOM or the EU after March 2019 [18].

Depending on the outcome of negotiations there may or may not be a transitionary period between March 2019 and December 2020, this period will allow the UK to operate under its current arrangements with the EU and EURATOM, whilst it implements its new arrangements.

Ultimately the immediate impacts of BREXIT on the UK domestic legislation for radiation protection are minimal, noting the regulatory regime was only recently updated to capture the latest EURATOM BSSD and ICRP recommendations (publication 103) [19]. Therefore, for all intents and purposes the UK's regulatory regime will meet best practice when the UK exits the EU, subject to some minor amendments to the regulatory process, to ensure the regulatory regime continues to function without involvement from the EU.

However, how the regulatory regime continues to remain in line with best practice in the future is not known at this stage. Through its membership with EURATOM the UK have always been an active contributor to the development of the EURATOM BSSD. However, once the UK ceases to be a member, the UK will no longer need to ensure compliance with the EURATOM BSSD, but rather the IAEA BSS.

Although the differences are fairly minor between the EURATOM BSSD and IAEA BSS, there is an increased importance in the UK relationship with the IAEA and ICRP to ensure it supports the development of the future iterations of IAEA BSS and feeds into the development of best practice. In addition, there needs to be commitment from the UK government to keep the UK radiation protection regulatory regime in line with the best practice.

There are also a number of wider potential impacts associated with BREXIT related to the supply, ownership and management of nuclear materials, such as nuclear fuel, and management of spent fuel and solid radioactive waste, however the impacts to management of radioactive discharges are minor.

9.7 Summary

- As a member state under the EURATOM treaty it is mandated that the EURATOM BSSD is incorporated into UK domestic legislation.
- The UK's domestic legislation for radiation protection (including the control of radioactive discharges) has recently been updated to incorporate the latest EURATOM BSSD and ICRP recommendations (publication 103).
- The key regulations controlling radioactive discharges in the UK are as follows:
    - **Environmental permitting regulations 2016 (as amended)** and **environmental authorisations (Scotland) regulations 2018**
        - Requires operators of certain facilities that could harm the environment or human health to obtain a 'permit' or 'permits' to allow them to undertake certain activities associated with the use of certain substances and the generation and disposal of waste.
        - Regulates the planned discharge of radioactivity from nuclear and non-nuclear establishments.
        - Additionally, covers the storage and use of radioactive materials on non-nuclear establishments.
        - Defines radioactivity limits for solid materials and waste below which the material/waste does not fall under the legal definition of radioactive (out of scope), but above which is does.
        - As part of the permit application the operator must demonstrate how it applies the principle of 'best available techniques', propose limits to any discharges of radioactivity, assess the potential radiological impacts of these discharges to humans and the

environment (including the flora and fauna), and describe how it will monitor these discharges and the environmental impacts.
    - Only once the permit is awarded and any relevant conditions met will the operator be allowed to begin undertaking the activities associated with their permit.
    - Throughout the operation of the facility and validity of the permit the operator will be required to meet the conditions of the permit including the need to continue to apply the principle of BAT, a requirement to monitor any waste arisings and discharges and periodically report these to the regulator and notify the regulator should they breach any limits or conditions of the permit.
- **Nuclear installations act 1965**
    - Regulates the activities undertaken on a nuclear establishment including the storage and use of radioactive materials and discharges of radioactive material under unplanned events.
    - Prior to the construction and operation of a nuclear establishment a corporate body is granted a nuclear site licence (NSL) to use a site for specified activities.
    - The NSL is a legal document, issued for the full life-cycle of the facility and defines a set of 36 standard conditions, covering design, construction, operation and decommissioning.
    - The licensee is required to demonstrate to the office for nuclear regulation (the regulator) throughout the various stages of the design, construction, commissioning, operation and decommissioning of the establishment, how all risks associated with the establishment are reduced to so far as is reasonably practicable (SFAIRP, the UK equivalent to the ALARA principle). This includes identification of those scenarios resulting in an accidental release of radioactivity into the environment, and how the design and operation of the establishment adequately prevents or mitigates against the impacts to the environment and the public.
- **Ionising radiations regulations 2017**
    - Regulates the use of radioactive materials and artificial sources to ensure that adequate protection is in place for individuals working with radiation and those that could be impacted by its use (e.g. visitors, members of the public) and ensure that radiation exposures are restricted so far as is reasonably practicable (SFAIRP).
    - The ionising radiations regulations are enforced by the health and safety executive (HSE) for non-nuclear establishments and office for nuclear regulation (ONR) for nuclear licensed sites.
    - One of the key requirements of the regulations is to conduct a radiation risk assessment.
    - The ionising radiations regulations also set out the legal dose limits for workers and members of the public in line with the IAEA and EURATOM basic safety standards and ICRP publication 103.

- **Radiation (emergency preparedness and public information) regulations 2001**
  - Establishes a framework to ensure that members of the public are properly informed and prepared in advance about what to do in a radiation emergency and provided with information should an event occur.
  - Relates specifically to an 'off-site emergency', namely where there is a large release of radioactivity into the environment which results, or is likely to result, in the need to consider urgent countermeasures to protect the public outside the site from the radiological hazard.
  - The regulations are enforced by the health and safety executive and establish a framework of roles and responsibilities for local responders including the duty to assess the risk of an emergency occurring and to maintain plans for the purposes of responding to an emergency.

- On 29 March 2017 the UK Prime Minister informed the European Union (EU) of the UK's intention to withdraw from both the European Union and the European Atomic Energy Community (EURATOM). Ultimately the immediate impacts of BREXIT to the UK's domestic legislation for radiation protection are minimal, noting the regulatory regime was only recently updated to capture the latest EURATOM BSSD and ICRP recommendations (publication 103). Therefore, for all intents and purposes the UK's regulatory regime will meet best practice when the UK exits the EU.

References

[1] ICRP 2007 The 2007 recommendations of the international commission on radiological protection ICRP publication 103 *Ann. ICRP* **37**

[2] IAEA 2014 *Radiation Protection and Safety of Radiation Sources: International Basic Safety Standards General Safety Requirements Part 3*

[3] European Union 2016 *The Euratom Treaty Consolidated Version* http://consilium.europa.eu/media/29775/qc0115106enn.pdf [accessed 2 September 2018]

[4] European Union 2013 *Council Directive 2013/59/Euratom of 5 December 2013 laying down basic safety standards for protection against the dangers arising from exposure to ionising radiation, and repealing Directives 89/618/Euratom, 90/641/Euratom, 96/29/Euratom, 97/43/Euratom and 2003/122/Euratom*

[5] Martin A, Harbison S, Beach K and Cole P 2012 *An Introduction to Radiation Protection* 6th edn (Boca Raton, FL: CRC Press)

[6] *The Environmental Permitting (England and Wales) Regulations 2016* No. 1154

[7] Department for Environment Food & Rural Affairs 2013 *Environmental Permitting Guidance Core Guidance for the Environmental Permitting (England and Wales) Regulations* 2010

[8] Department for Environment Food & Rural Affairs 2011 *Environmental Permitting Guidance Radioactive Substances Regulations 2010 Version 2.0*

[9] Nuclear Industry Safety Directors' Forum 2010 *Best Available Techniques (BAT) for the Management of the Generation and Disposal of Radioactive Wastes - A Nuclear Industry Code of Practice*

[10] Environment Agency 2012 *Principles for the Assessment of Prospective Public Doses arising from Authorised Discharges of Radioactive Waste to the Environment*

[11] *The Environmental Authorisations (Scotland) Regulations 2018* No. 219

[12] *Nuclear Installations Act 1965* c.57

[13] Office for Nuclear Regulation 2017 *Licence Condition Handbook*

[14] *The Ionising Radiations Regulations 2017* No. 1075

[15] Health and Safety Executive 2017 *Work with Ionising Radiation Approved Code of Practice and Guidance* L121

[16] *The Radiation (Emergency Preparedness and Public Information) Regulations 2001* No. 2975

[17] Health and Safety Executive 2001 *A guide to the Radiation (Emergency Preparedness and Public Information) Regulations 2001* L126

[18] Department for Business Energy and Industrial Strategy 2018 *Guidance - Civil Nuclear Regulation If There's No BREXIT Deal*

[19] Barrett J, Broughton J, Bryant P, Clark S, Perks C, Thurston J and Webbe-Wood D 2018 *Exit from the EU and the EURATOM Treaty: The Implications for Radiation Protection in the UK*

Lightning Source UK Ltd.
Milton Keynes UK
UKHW051418210920
370123UK00009B/67